Transportation Engineering: Planning and Design

Transportation Engineering: Planning and Design

Cooper Hunt

CLANRYE INTERNATIONAL
www.clanryeinternational.com

Clanrye International,
750 Third Avenue, 9th Floor,
New York, NY 10017, USA

ISBN: 978-1-64726-644-8

Cataloging-in-Publication Data

Transportation engineering : planning and design / Cooper Hunt.
p. cm.
Includes bibliographical references and index.
ISBN 978-1-64726-644-8
1. Transportation engineering. 2. Civil engineering. 3. Transportation. I. Hunt, Cooper.
TA1145 .T73 2023
629.04--dc23

For information on all Clanrye International publications
visit our website at www.clanryeinternational.com

CLANRYE
INTERNATIONAL

Contents

Preface

This book has been an outcome of determined endeavour from a group of educationists in the field. The primary objective was to involve a broad spectrum of professionals from diverse cultural background involved in the field for developing new researches. The book not only targets students but also scholars pursuing higher research for further enhancement of the theoretical and practical applications of the subject.

Transportation engineering refers to the use of scientific principles and technology to operate, plan, manage and create functional design of facilities for different modes of transportation. It aims to provide comfortable, economical, safe, rapid, convenient, efficient and environment friendly transportation for people and goods. It is involved in the planning, designing, building, maintaining and operating transportation facilities, which facilitate railroad, water, air, pipeline, highway and space transportation. The designing aspects of transportation engineering encompass determining the size of transportation facilities, and identifying the materials and thickness utilized in pavement design of the roadway. The planning aspects in transportation engineering are connected to urban planning and entail technical forecasting decisions as well as political considerations. The technical forecast of passenger travel typically contains an urban transportation planning model, which involves the assessment of mode choice, trip generation, route assignment and trip distribution. This book elucidates the planning and design aspects with respect to transportation engineering. It aims to serve as a resource guide for students and experts alike and contribute to the growth of the discipline.

It was an honour to edit such a profound book and also a challenging task to compile and examine all the relevant data for accuracy and originality. I wish to acknowledge the efforts of the contributors for submitting such brilliant and diverse chapters in the field and for endlessly working for the completion of the book. Last, but not the least; I thank my family for being a constant source of support in all my research endeavours.

Cooper Hunt

1

Highway Planning and Alignment

1.1 Highway Development in India

Existence of roads in India dates back to the remote past. Highway Development in India from ancient age to modern age are briefly discussed below:

Before BC

Excavations of cities like Mohanjodaro and Harappa which existed between 3500 BC to 2500 BC have revealed the existence of broad streets. It is found that the importance of roads were realized particularly for military and administrative purposes.

Ancient scripts reveals that a road of 6 to 7.5 m wide was built by King Bimbasara in 600 BC. Ancient literature known as Artha Sasthra furnishes adequate information on the road construction during the Aryan period.

Kautilya, the first Prime Minister of Chandra Gupta Maurya constructed a highway connecting North West Frontier province to Patna during 300 BC. During 270 BC King Ashoka constructed a good network of roads.

Mughal Period

Road development received special attention during this period. In this period twenty four long roads connecting different parts of the country were constructed. Out of which thirteen were traced. During this period roads were greatly improved and well maintained. Mughals received great appreciation from the foreign visitors who visited India during the region.

British Period

At the beginning of British rule, the condition of roads in India were very much damaged. Economic and political shifts caused deterioration to road transportation. The fall of Mughal's rule led to the ruin of the road communication system.

Due to military movement, britishers got a few truck roads metalled and got bridges constructed. Only old roads were reparied or metalled. Lord Dalhousie, the Governor General of India formed the Central Public Works Department to look after the roads in India.

The construction of Grand Trunk Road was undertaken by this new Department. Lord William Bentick, Lord Mayo and Lord Ripon contributed measurably to the road construction in India. The introduction of railways in 1853 was a serious set back faced by road construction.

Only a few road constructions of military importance was carried out. After World War I, a revolution in India's transport system came due to the cheap supply of surplus military vehicles.

In order to bridge the traffic due to motor vehicles and village bullocks, refinements were needed in the practice of road construction. Under this condition Jayakar Committee was formed, recommendations of the committee were implemented in phased manner.

During the World War II (1939), the roads were under the control of provinces and due to lack of funds, the provinces could not meet the requirements of road development. After the war period, it was realized that the effective system of roads could be active only if it is under central government. It was felt if not all roads at least a few of them could be maintained by the central government.

A conference of chief engineers were held at Nagpur in 1943 which made very useful recommendations. Conference has been commonly known as Nagpur Plan.

Jayakar Committee Recommendations and Realisations

After the World War I, number of motor vehicles gradually increased and demanded a better network of roads. There was a need for the roads to carry both bullock cart traffic and motor vehicles.

But existing roads were not laid to withstand the mixed traffic conditions. In 1927, the Indian Roads and Transport Development Association (IRTDA) was set up having head office at Bombay and branch offices at Delhi, Gauhati and Madras (Chennai).

It's function was to study the transport problems of the country and recommend improvements. Mr. M.R. Jayakar was appointed as the Chairman by the Central Government for a committee called as "Jayakar Committee".

The committee looked into the road development in India and submitted a report based on the recommendations of IRTDA. The Jayakar Committee submitted its report in 1928.

Road development should be considered as a national problem and necessary attention should be paid. To meet the road development expenses, an extra tax to be levied on petrol and to be named as central road fund.

An advisory board on road development has to be constituted with experts pooled from different sectors who have knowledge on road development. A research organisation to be set up at national level to undertake research on roads and road related areas and provide consultancy services to the Government as and when needed.

As per the recommendations,"Central Road Fund" was created on 18th March 1929. The consumers of petrol were then charged an extra levy of 2.04 paise per litre (i.e., two annas per gallon).

Twenty percent of the revenue collected through the fund was retained as central reserve and the balance allotted to the various states based on actual petrol consumption. Based on the recommendation, a semi-official technical committee known as Indian Roads Congress (IRC) was formed in 1934.

Major Roles of IRC:

- To promote and encourage the science and practice of construction and maintenance of roads.
- To suggest improved methods of administration, planning, design, construction, operation, use and maintenance of roads.
- To promote the use of standard specifications and to propose specifications.
- To advise regarding education, experience and research connected with roads.
- To suggest legislation for the development, improvement and protection of roads.
- To publish or arrange for the publication of proceedings, journals, periodicals and other literature for the promotion of the objects of the society.
- Based on the recommendations of the Jayakar Committee in the year 1950, a research institute exclusively for research on roads was started under the name, Central Road Research Institute (CRRI) at New Delhi.

Functions of CRRI

Central Road Research Institute is engaged in carrying out the following functions:

1. Research and development projects on design, construction and maintenance of roads.
2. Traffic and transportation planning of mega and medium cities.
3. Management of roads in different terrains.
4. Improvement of marginal materials and utilization of industrial waste in road construction.
5. Landslide control and ground improvements.
6. Environmental pollution.
7. Road traffic safety analysis and design.

8. Service life assessment and rehabilitation of highway and railway bridges.
9. Performance monitoring and evaluation of roads.
10. Technical and consultancy services to various user organisations.

1.2 Classification of Roads

Roads are generally classified based on the following factors:

Traffic Volume

Classification based on traffic volume or tonnage have been fixed by different agencies arbitrarily hence it may not be agreeable to all.

Based on traffic volume, roads may be classified as heavy, medium and light traffic roads. These terms can be expressed as number of vehicles per day.

Load Transported or Tonnage

Load or tonnage is also relative and roads may be classified as :

Class I, Class II, Class III, etc. and the limits may be expressed as tonnes per day.

Location and Function

The classification based on location and function has been accepted by every one.

Accordingly the Nagpur Road Plan classified the roads in India into five categories, viz.,

i. National Highways (NH)
ii. State Highways (SH)
iii. Major District Roads (MDR)
iv. Other District Roads (ODR)
v. Village Roads (VR)

Speed of Vehicles

These classifications have been accepted fully and has been functioning over the years. The classification based on speed of a vehicle is new one which is not fully defined.

However, roads identified to allow vehicles to go with a minimum speed have been classified as expressways.

National Highways

These are the main highways running through the length and breadth of the country. It connects ports, foreign highways, capitals of large states and including roads required for strategic movements for the defense of India.

For underdeveloped areas it was proposed to construct national traits so that at a later stage they could be connected into appropriate roads to suit the traffic requirements.

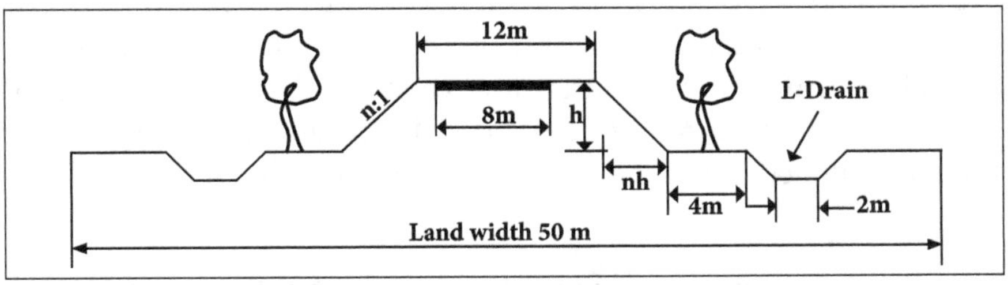

Typical cross-section of National Highway.

National highways should form the framework of roads on which the entire road construction of the country should be based. Selection of National highways should be such that they afford uninterrupted road communication.

It is not mandatory that all National Highways should be same type and dimensions. All the National Highways are assigned a specific number.

Example: Highway connecting Delhi-Chandigarh-Amristar is denoted as NH-1.

Construction and maintenance of the National Highways rest with the Central Government Departments such as Central Public Works Department.

Typical cross-section of a National Highway is shown in figure. National Highways are of two lanes of atleast 8 m width and atleast 2 m wide on either side.

Provincial or State Highways

These are all other main trunk or arterial roads of a province or state, connecting up with the National Highways of adjacent states, district head quarters and important cities within the state. It serves as the main arteries of traffic to and from district roads.

These are main roads of commercial interest to move goods within a state.

They may sometimes allow heavier traffic than National Highways and for this sake its designation cannot be changed.

These roads are of two lanes and have similar cross-section as that of National Highways. Construction and maintenance are handled by State Government Public Works Department. Central Government grants are provided for the development of State Highways.

District Roads

These are those roads which are traversing each district serving areas of production and markets. It connects the district serving areas of production and markets with each other or with highways and railways. They take road traffic into the main parts of rural areas through the year with only minor interruption. It is built and maintained by local self government.

Every village in a highly populated area should be within 3.2 km of a district road. These roads should form main branches from the National or State Highways, should be capable of taking traffic into the interior of each district or similar unit of the area. Figure shows a typical urban road of a district. With the view of expected traffic and the type of finish provided, district roads were further divided into two classes as given below:

Major District Roads: It connect all important towns of a district with head quarters. Also connects the areas of production, market, railway stations and State Highways. They roughly have the same specification as that of State Highways.

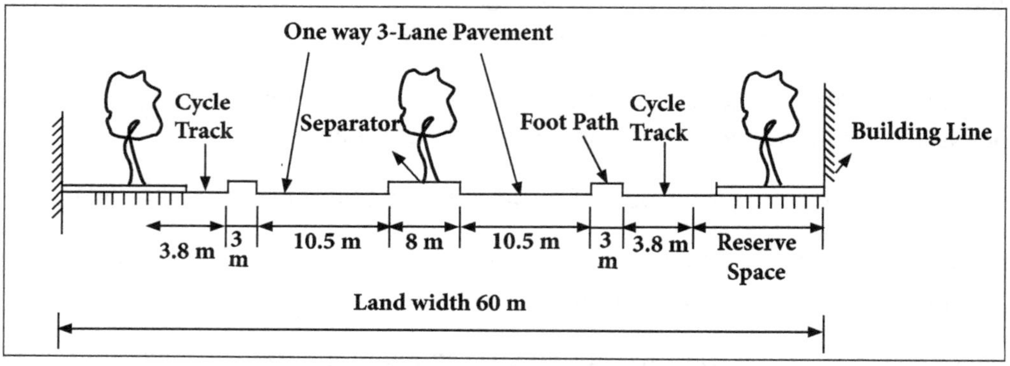

Typical cross-section of urban road.

Other District Roads: These are also called as Minor District Roads. They connect a town with other towns or important villages of a district. They also connect other roads with State Highways. These are of somewhat lower specifications than the above types. During the monsoon periods many of these roads are kept closed. To be used throughout the year, necessary drains should be made. These are built and maintained by Municipal Boards and Corporations.

Village Roads

These are the roads which are connecting villages and groups of villages with each other and the nearest district road, main highway, railway or river ghat. These roads are unmetalled, connecting villages together or a village to a railway station or district road.

The village tracks should not be left to develop by itself but should be replaced by a plan for a designed and regulated system. Local District Boards are responsible for the construction and maintenance of village roads. Typical cross section of a village road is shown in figure.

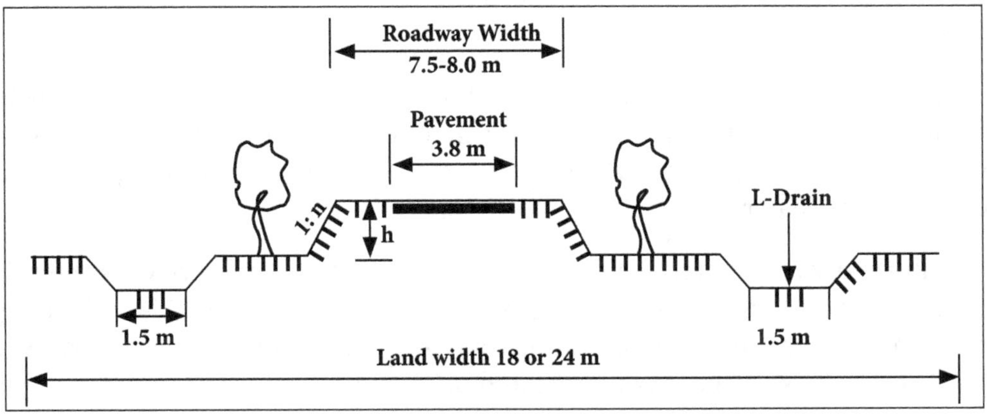

Typical cross-section of a village road.

1.3 Road Network Patterns

Roadway patterns are very essential in the development of the settlements of a city. But, recent development in the cities does not give importance to the study of the road patterns that give rise to numerous roads which are not interconnected, housing schemes and commercial developments built far away where the roads are very distant from the center of the town.

Then, by increasing distance between the residential and commercial hub of the city increases the dependency upon cars for the daily travel chores each household member makes frequently. Also, the roadway patterns increase the response time the emergency response vehicles take to reach a certain place.

Types of Road Patterns

The road network can be laid in different patterns. They are:

- Rectangular or Block Pattern
- Radial or Star and Circular Pattern
- Radial or Star and Block Pattern
- Radial or Star and Grid Pattern
- Minimum Travel Pattern
- Hexagonal Pattern

1. Rectangular or Block or Grid Pattern

The rectangular or block or grid pattern is a plan where the streets and roads are in the form of grids or blocks that is running perpendicularly into each other hence forming a grid or block.

For example: The North Nazimabad area of Karachi in Pakistan where the streets and roads run at an angle of ninety degree into each other. Generally, in this type of pattern, the streets and roads are far away from each other so that it takes a long time to reach the center of the area or city.

They are normally considered weak from the road safety point of view as the vehicles meet at opposite directions at intersections. Generally, the main road is wide that passes through the centre of the area and other roads connecting to it are narrow. Such pattern is sufficiently easier to construct and maintain.

2. Radial Pattern

Radial pattern is the network of roads which is in the form of circles emanating from the centre of the area. For example: The paradigm of radial pattern can be found at the Federal B Area of Karachi.

It is a combination of radial and block pattern. The radial network of roads radiate from the center outwardly with the block pattern network of roads in between the radial main streets. This pattern is represented in the below figure:

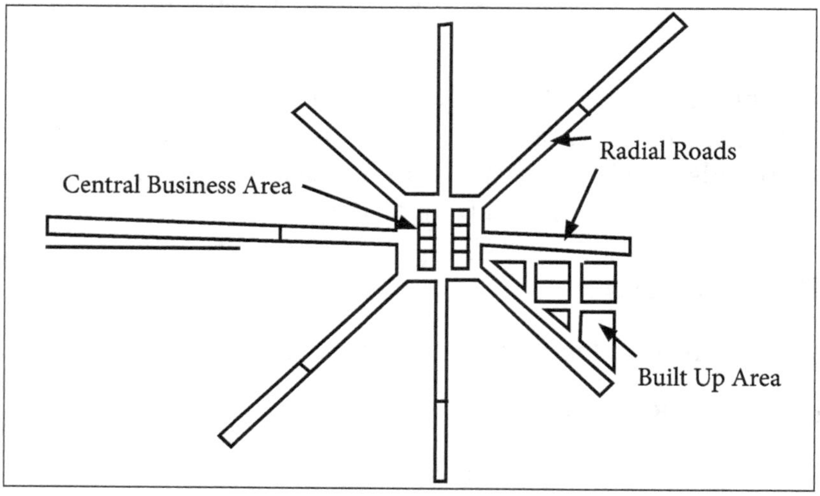

Radial and block pattern.

3. Radial and Grid Pattern

This is a combination of radial and grid pattern. The radial network of roads radiate from the center outwardly. Therefore, the main radial streets are interconnected by providing grid pattern in between the main streets. This pattern is represented in the figure:

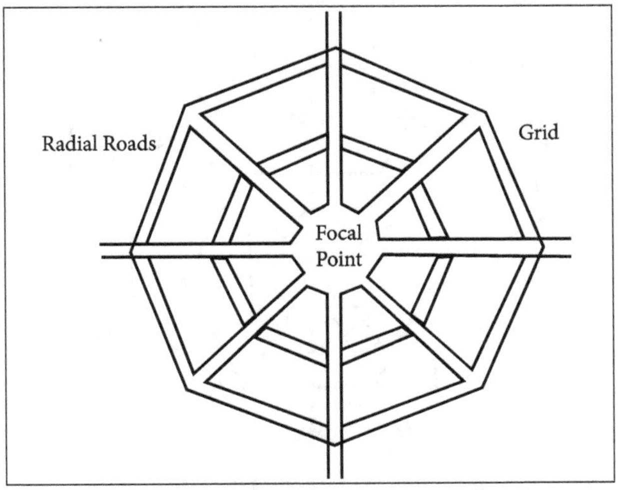

Radial and grid pattern.

4. Radial and Circular Pattern

This is a combination of radial and circular pattern. The main roads radiate from the central point and they are connected together by concentric roads which are also radiating outwardly.

5. Hexagonal Pattern

It is a network of roads which grow in various directions forming hexagons as shown in the below figure.

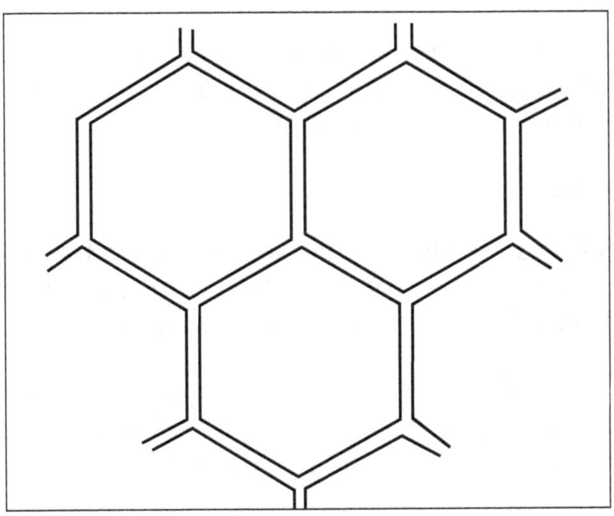

Hexagonal pattern

6. Linear Pattern

The linear patterns are the roads that grows linearly in one direction possibly because of the presence of some natural forces such as sea or ocean at one side of the city.

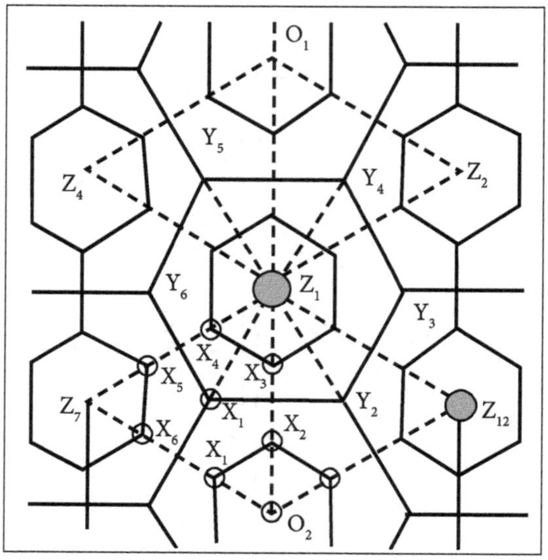

Minimum travel pattern.

In such road pattern, the city is contented by suburban center, sector center and neighborhood center by the road that required minimum to connect the city center.

1.3.1 Necessity for Highway Planning

Significance of Transport

The transport plays a major role in the overall economic development. The transportation results into growth of industrialization, infrastructure and large production.

Comfort and convenience has been obtained from the advancement in the transport sector. The well-functioning transportation systems form the basis for the economic prosperity as well as social well being of societies.

1. Industrial Growth

Generally, transportation and industrial development are interconnected. Without improved modes of transportation it might have been harder for the industrial producers for producing as well as selling their goods to the wider markets.

Therefore, transportation facilitates movement of raw material and different requirement from the place of supply to the place of production. An economical transport is indispensable to the economic development of the nation.

2. Creates Employment

Additionally, it contributes to the economic development through job creation. This creates both direct and indirect employment opportunities. Transportation also facilitates movement of labors so that it encourages employment resulting into industrial development by improving economic development.

In India, a sizeable portion of the country's working population is either directly or indirectly employed in the transport sector.

3. Creates Place Utility

It permits movement of commodities from the producer to the final consumer at any time or place when they are demanded. This creates place utility. Thus, it plays an essential role in the agricultural sector.

The agricultural needs are made available to the farmer at a short span of time where it becomes an integral part of commerce. This also gives place and time utility to goods by removing them from the place of production to the places where they are used.

4. Bring Countries Closer

Not every country in the world is self-sufficient. It needs to depend on one another in order to fulfill their needs. But, transportation has brought the countries closer. This not only caters to the requirement of mobility even it provides comfort and convenience.

Nowadays, traveling has become a part of our daily lives where people travel for education purpose, business purpose and vacation purpose and so on. The transport system is doing a great job by easing the pain of covering a vast distance of land which help in bringing the countries closer.

5. Serve Several Purposes

It provides access to the natural resources and promotes trade, by permitting a nation to accumulate wealth and power. Also, it allows the movement of equipment, soldiers and supplies during war.

Thus, it has become important to a nation's economy since it serve several purposes. This includes the production and distribution of fuel, the manufacture and distribution of vehicles and the provision of transportation services.

6. Stability in Prices

The goods shall be transported to places where scarcity is present and the prices are high from places where there is surplus and the costs are low. This transfer of products from the place of surplus to the place of scarcity will enable to stabilise the prices of the commodity. Hence, the stability of prices restricts the local producers for charging prices at their own will where it shall discourage the monopoly and encourage competition.

7. Specialization and Division of Labour

It will increase the mobility of labour and capital widens the market which results in specialization and division of labour that helps in stabilizing prices. The specialization provides employment to a very large number of persons.

This is because of only transport in which the modern industrial system and large-scale industries are in a position to develop. Without the help of efficient transport it might not be possible to procure the raw material, gather large number of workers and distribute the finished goods.

8. Use of Economic Resources

The transportation permits society to enjoy advantages of specializations of resources and the advantages of the labour by making it available for products which is to be brought great distance hence avoiding the necessity for local production for all conceivable commodities of need.

Thus, every economic region will concentrate upon the goods and services for which is best adapted either through the natural resources endowment or through the historical development. Therefore, it helps to lead to a better economic use of available resources.

9. Standard of Living

Transportation raises the quality of living, creating possible improved housing, clothing, food and recreation.

Significance of Highway Planning

For development of highways, a proper highway planning is needed. Every highway planning is done keeping in mind the following significance:

1. To establish an integrated highway network with minimum cost capable of encompassing all highway travel in a most safe, efficient, economical and orderly manner. The minimum cost should include initial cost of construction, maintenance cost, renewal cost of pavement layers and the vehicle operation costs.
2. To plan a highway system which could be constructed within the available resources during the proposed plan period with maximum utility.
3. To forecast the future requirements of road kilometer needed and improvement of roads.
4. To phase out the entire highway development during the plan period by identifying the priorities keeping in view,the utility as the main criterion.
5. To work out financing system.

1.4 First, Second and Third Road Development Plans

Road transport in the modern sense that is the vehicles are driven by internal combustion engines by using petrol or diesel as fuel where it was practically negligible in India before the World War II. The following plans have been drawn to develop the roadways in India.

Twenty Year Plan

After achieving the objectives of the Nagpur Plan, another plan called as Twenty Year Road Plan was drawn in 1961. This actually aimed at increasing the road length from 6.56 lakh km to 10.60 lakh km and the density to 32 km of the road per 100 sq. km by 1981.

The remaining objectives of the Twenty Year Road Plan were:

i. To bring every village in the semi-developed area within 12.8 km of a metalled road.

ii. To bring every village in an undeveloped and uncultivated area within 19.2 km of a metalled road and 8 km of any other road.

iii. To bring every village in the developed agricultural area within 6.4 km of a metalled road and 2.4 km of any other road.

Road Development in India

At the Nagpur Conference in 1943, the first attempt for the proper planning of the highway development programme in India on a long term basis was made. Once the completion of the Nagpur Road Plan, the Second Twenty year Plan was drawn for the period of 1961-1981. Then the Third Twenty Year Road Development Plan was drawn for the period of 1981-2001 was approved only during the year 1984.

First 20-Year Road Plan

It was formed in the year 1943 at Nagpur and the plan period was from 1943 to 1963. The two plan formulae were finalized at the Nagpur Conference to decide two categories of road length for the country as a whole and also for individual areas. It was the first attempt for highway planning in India. Therefore, the two plan formulae assumed the Star and Grid pattern of road network. Thus, the two formulae are also called "Star and Grid Formulae".

It is envisaged increasing of the kilometreage of major roads to 1,96,800 km and of other roads to 3, 32,800 km by 1953. In this case, the highlight of the plan was that no village in a developed agricultural region must be more than 8 km from a major road or 3 km away from any other road whereas the average distance of villages from a major road must be less than 3.2 km.

In case of a non-agricultural region, these distances were fixed at 32, 8 and 10 km respectively. Such plan could not be implemented immediately since the country was ruled by a number of princely states outside British India. Therefore the concerted efforts to achieve the objectives of this plan were made only after the re-organisation of the states. Then the targets of this plan were more or less achieved by 1961.

The salient features of Nagpur Road Plan are as follows:

The roads are classified as:

1. Primary System (NH/Expressways)
2. Secondary System (SH/MDR)
3. Tertiary System(ODR/VR)

The construction and maintenance of NH was assigned to Central government. This actually aims for 2, 00,000km of bituminous road and 3, 32,700km of other types. Hence, the formulae was based on star and grid pattern of road network and was considered as two equations:

$$\text{NH+SH+MDR(km)} = \left[\frac{A}{8} + \frac{B}{32} + 1.6N + 8T + D - R\right]$$

I -Category: NH/SH/MDR

where,

A = Agricultural area, km^2

B = Non-agricultural area, km^2

N = Number of towns and villages with population range 2001-5000.

T = Number of towns and villages with population over 5000.

D = Development allowance of 15 percent of road length calculated to be provided for agricultural and Industrial development during the next 20 years.

R = Existing length of railway track, km.

II-Category: ODR/VR

The total length of second category roads for other district road and village roads in km is given by the formula:

$$\text{ODR + VR (km)} = [0.32\,V + 0.8Q + 1.6P + 3.2S] + D$$

where,

V = Number of villages with population 500 or less.

Q = Number of villages with population range 501-1000.

P = Number of villages with population range 1001-2000.

S = Number of villages with population rang 2001-5000.

D = Development allowance of 15 % for next 20 years.

NH/SH/MDR are meant to provide the main grids and ODR/VR as the internal road system. Therefore the development allowance is 15% and the length of railway track was deducted.

Second Twenty Year Road Plan

Since the target road length of Nagpur road plan was completed nearly earlier in 1961 a long term plan was initiated for twenty year period which was initiated by IRC. Thus, the second twenty year road plan came into picture which was drawn for the period of 1961-81. Therefore the second twenty year road plan was envisaged overall road length of 10, 57,330 km by the year 1981.

The salient features of second 20 year road plan are as follows:

Every town with population above 2000 in plains, 500 in hilly areas and 1000 in semi-hilly areas must be connected by a bituminous road or metaled road.

It aims to provide 32km/100 sq. km area. 1600 km length of expressways was proposed. The development allowance is 5% only and the length of railway track was not deducted. The five equations are given to find NH/SH/MDR/ODR/VR.

i. National Highways (km)

$$=\left[\frac{A}{64}+\frac{B}{80}+\frac{C}{96}\right]+\left[32K+8M\right]+D$$

ii. National Highways + State Highways (km)

$$=\left[\frac{A}{20}+\frac{B}{24}+\frac{C}{32}\right]+\left[48K+24M+11.2N+1.6P\right]+D$$

iii. National Highways + State Highways +Major District Roads (km)

$$=\left[\frac{A}{8}+\frac{B}{16}+\frac{C}{24}\right]+\left[48K+24M+11.2N+9.6P+6.4Q+2.4R\right]+D$$

iv. National Highways + State Highways +Major District Roads + Other District Roads (km)

$$=\left[\frac{3A}{16}+\frac{B}{32}+\frac{C}{16}\right]+\left[48K+24M+11.2N+9.6P+12.8Q+4R+0.8S+0.32T\right]+D$$

v. National Highways + State Highways +Major District Roads + Other District Roads + Village Roads i.e., all roads (km)

$$=\left[\frac{A}{4}+\frac{B}{8}+\frac{C}{12}\right]+\left[48K+24M+11.2N+9.6P+12.8Q+5.9R+1.6S+0.64T+0.2V\right]+D$$

where,

A = Developed and agricultural areas, km^2

B = Semi-developed area, km^2

C = Undeveloped area, km^2

D = Development allowance of 5 percent of road length calculated for further development and other unforeseen factors.

K = Number of towns with population over 1,00,000

M = Number of towns with population range 1,00,000-50,000

N = Number of towns with population range 50,000-20,000

P = Number of towns with population range 20,000-10,000

Q = Number of towns with population range 10,000-5,000

R = Number of towns with population range 5,000-2,000

S = Number of towns with population range 2,000-1,000

T = Number of towns with population range 1,000-500

V = Number of towns with range below 500

Table: Comparison of I and II 20-year plans.

First 20-year Plan	Second 20-year Plan
Two equations are given to find NH/SH/MDR/ODR/VR.	Five equations are given to find NH/SH/MDR/ODR/VR.
Areas is divided into two types namely, developed and semi-developed.	Areas is divided into three types namely, developed areas developed, semi-developed and undeveloped area.
The development allowance is 15%.	The development allowance is 5%.
No express ways.	1600km of expressway was included.
It aims to provide 16km/100 sqkm.	It aims to provide 32km/100 sqkm.
Length of Railway track was deducted.	Not deducted.

Third Twenty Year Road Plan

The third twenty year road plan was prepared by the Road Wing of the Ministry of Shipping and the Transport with the active co-operation from a number of organizations and the experts in the field of Highway Engineering and Transportation. Such document was released during the 45th Annual Session and in the Golden Jubilee celebrations of the Indian Road Congress in February 1985 at Lucknow. Hence, this plan for 1981-2001 is also known as 'Lucknow Road Plan'.

The road development is based on the primary, secondary and tertiary road system. So, all the villages with population above 500 is to be connected with the metalled road. It aims to provide 82 km/100 sq. km. The NH will form the main guideline. Then the length of SH and MDR are calculated based on the area of population. The long term master plans must be prepared at taluk, district and state levels.

The salient features of second twenty year road plan are as follows:

The roads must also be built in less industrialized area and the existing roads must be improved. Additionally, there must be improved environmental quality and road safety.

1.4.1 Rural Road Development Plan

The Rural Development Plan includes construction of rural roads under Rural Landless Employment Guarantee Programme (RLEGP), Minimum Needs Programme (MNP), Command Area Development (CAD) programmes and Jawahar Rojgar Yojana (JRY) to connect all the villages having a population of 1,500 or more with all weather roads and those having less than 1,500 population with link roads.

Build Operate Transfer (BOT): The build operate transfer is a scheme under which the private operators are invited to construct roads and bridges. Generally, they are allowed to collect the toll tax from the vehicles using these roads and bridges for a specific period of time until these assets are transferred to the government. The National Highways Act was amended to facilitate the private investment in the real construction under the BOT scheme.

Central Road Fund (CRF): The CRF is being raised for the betterment of the roads by imposing additional excise or customs duty at the rate of Rs. 1.50 per litre on the petrol with an effect from 2 June 1998 and on the High Speed Diesel with an effect from February 28, 1999. Thus, the annual accrual through this sources shall be about Rs. 5,500 crore and a part from this shall go to fund the NHDP.

With the primary objective of providing regular and adequate flow of funds for the development of the road sector, the Central Road Fund Act 2000 was enacted in December, 2000. This is a non-lapsable fund.

The Act empowers the Centre to administer by managing and allocating the accrued amount to the following:

1. Development and maintenance of National Highways.
2. Development of rural roads: About 43 per cent of the levy on diesel will be spent on improving the rural connectivity.
3. Development and maintenance of State roads.
4. Construction of road under or over bridges and the safety works at unmanned railway crossings.

1.4.2 Vision 2025

The 'RURAL ROAD DEVELOPMENT PLAN: VISION 2025' was released on 23-May, 2007 by the Prime Minister, Dr. Manmohan Singh during the inaugural session of the two-day National Conference on Rural Roads organized by the organized by the National Rural Roads Development Agency (NRRDA), an autonomous agency under the Ministry of Rural Development and the Construction Industry Development Council (CIDC).

The Rural Roads Programme known as the Pradhan Mantri Gram Sadak Yojana (PMGSY) was launched in December 2000 as a 100 percent centrally funded scheme to give a boost to rural connectivity. This estimates that about 1.79 lakh unconnected habitations must be taken up under the PMGSY programme. So that, this shall involve a new construction in a length of about 375,000 km at an estimated cost of Rs. 78,000 crore and also improvements of 372,000 km at an estimated cost of Rs. 59,000 crore. Till the end of December, 2006, a total of about 83,000 habitations were covered and the rural road works for an amount of Rs. 38,387 crore was sanctioned.

Other than the balance work of PMGSY that is itself is an ambitions task, where it has a need to tackle the requirement of connectivity to habitations with lower population in the interest of social justice and removing sense of isolation of remote villages and bring them into main stream of national life. The objective has to be to provide full connectivity to all habitations including provision of bridges and culverts. Accordingly, the following vision for new connectivity has been recommended.

Habitations with population above 1000	Year 2009-10*
(500 in case of hill, NE states, deserts and tribal areas)	
Habitations with population above 500	Year 2014-15
(250 in case of hill, NE states, deserts and tribal areas)	
Habitations with population above 250	Year 2021-22

* These are the targets as per the Bharat Nirman Programme

The Central Government has also introduced the concept of a Core Network, which is defined as the network that is essential to provide one basic access to each habitation. The Core Network comprises of routes and link routes. Through these routes, it shall collect traffic from several link roads or a long chain of the habitations and it shall lead to a market centre or a higher category road. The link routes connects the single habitation or a group of habitations to the through routes or district roads leading to market centres. Similarly, through routes have the potential of getting upgraded as Major District Roads in future. As such, the route alignment, design and construction standards for such roads be decided accordingly.

Background

The government, as a part of its Poverty Reduction Strategy which launched a major

Rural Roads Connectivity Programme known as "Pradhan Mantri Gram Sadak Yojana (PMGSY)" in December 2000. When implementing this programme, several new initiatives taken by the Ministry of Rural Development, were expected to lead to the overall Sectoral Development, which includes the development of specifications, procurement and management standards and so on. By expanding on these initiatives, the Ministry had decided to prepare a Vision Document for the development of the Rural Roads in India on the similar lines of Vision 2021 for the Highways brought out by Indian Roads Congress (IRC) on behalf of the Ministry of Road Transport & Highways (MoRT&H).

Then the IRC also accepted the responsibility of developing the Document as well as constituted an Expert Group with Shri. D.P. Gupta, who was the former DG, MoRT&H as Convener along with Dr. L.R. Kadiyali and Shri. P.K. Lauria as Members.

Consultative Process for the Development of Vision 2025

The Expert Group appointed for the development of the Document came out with a lead paper outlining the content of the Document and presented it in the 173rd Council Meeting of IRC held on 9th January 2005 at Bangalore, for getting the views and opinions of the Highways Engineers across the country. This was followed by a Workshop on 'Integrated Development of Rural and Arterial Road Network: Rural Roads Vision 2025' organized jointly by IRC and NRRDA during 25th-27th February 2005 at Vigyan Bhawan, New Delhi.

Based on the inputs from the above Workshop, the Expert Group has fine-tuned the broad outline content of the proposed document and developed a Questionnaire for collecting data on Rural Roads Sector from the States.

Then the states were also asked to give information on the following issues:

- Investment patterns in Rural Roads Sector.
- Employment Generation in Rural Road Construction.
- Rural Transportation Systems.
- Construction Technologies adopted.
- Land Acquisition Policy.
- Norms followed for Rural Roads.
- Maintenance Policy.
- Role of Gram/Block/Zilla Panchayats.
- Technical Support available to PRIs.
- Status of Rural Roads Constructed under employment schemes.

- Status of Contracting Industry.
- Government-Academicia partnership
- Incorporation of rural roads network concepts in rural development planning.

While the data from more than 25 states has been received for analysis, the same is shortly expected from the rest of the states.

The MoRD has constituted a Committee for the Rural Roads Development Plan Vision 2025 with JS (RC) & DG, NRRDA as Convener and Secretary as well as IRC as a Member Secretary of the Committee. So, the members are drawn from different organizations connected with the road transport, Representatives of Consulting Firms, Representatives from the States who are dealing with Rural Roads, NCAER, NGOs, Ministry of Panchayati Raj and the two corresponding members from Academicia.

By continuing the action as well as in order to discuss further details about the scope and content on the Vision Document, a second workshop was organized at Vigyan Bhawan by NRRDA on 30th April 2005, New Delhi with the participation of the Members of the Committee constituted by the Ministry of Rural Development.

Then, a presentation on the proposed document and stressed the need for reliable data base in order to develop the Vision Document for assessing the needs and impacts was made by Shri. D.P. Gupta, Convener of the Expert Group. The several useful suggestions were put forward by the delegates that will also be considered in the development of the Vision Document.

Those suggestions which covered are the role and impact of Rural Roads, adoption of integrated rural accessibility planning, generations of alternative planning scenarios, development of need based designs, need for cost effective construction with the use of local materials, use of appropriate technology, issues in land acquisition and appropriate legislation to avoid litigation, quality assurance, stage construction, rural roads safety etc. Also, it was highlighted that the development of Rural Roads must be linked to Millennium Development Goals (MDGs) at local and global level.

Then the delegates stressed the need for coordination among the different agencies and programmes aw well as the steps that must be taken for asset management. Also, Rural Roads should be viewed as a 'means' rather than the 'end' in any employment generation strategy.

They suggested that the Vision Document must clearly give a word of caution on the depletion of natural resources and the need for appropriate policies in the construction technology. Additionally, another important issue which came to light is the role of Panchayati Raj Institutions (PRIs) and the ways and means of community participation in asset creation and management, with the necessary capacity building. Therefore, it suggests the document must clearly bring out the policies on the resource mobilization and fiscal planning.

Also, the deliberations pointed out the need for identifying the key transport indicators for the bench marking and evaluation of the Rural Roads Sectoral Programmes.

Current Status

Here, the Expert Committee has to prepare and circulate a Draft Vision Document which should be incorporating the suggestions and views from the two workshops that was held earlier and the database received from the States. Such document will be discussed by the Steering Committee appointed by IRC and later in the 175th Council Meeting of IRC which was held on 22nd May 2005 at Shillong and also more suggestions for fine-tuning and finalizing the document was received from the Members.

The Expert Committee is expected to submit the Draft of the Final Report of the Vision Document by 15th June 2005. The IRC shall finalize the document after the approval by the Ministry of Rural Development and later the approved version of the Rural Roads Development Plan: Vision-2025 by middle of July 2005was published.

1.5 Planning Surveys

Generally, highway surveying is a specialized type of land surveying that is conducted for government agencies. It is often conducted at the time of highway planning stages. When the construction project is carried, a highway construction survey is used to ensure that the process is going to the procedure it has to be as well as it is located precisely where it belongs. Once the highway is built, a highway survey may be used for providing an accurate layout of roadways, storm drainage systems, utilities, nearby buildings, overhead wires and other features of the landscape.

Often, the construction surveyor is involved in the placement process. It actually requires a survey of the existing terrain in order to find the easiest route. Therefore, highway construction surveys are particularly concerned with areas where the dirt must be moved, noting to what level and grade. The main goal of such survey is to determine the appropriate route in which the least amount of land that needs to be moved. Thus, an engineer may use the survey data for determining where the the dirt should be moved in order to reduce the total amount of dirt relocation.

Though do surveyors must determine how the land should be prepared before the highway can be constructed, however they also assist engineers in the mapping of optimal routes. It not only includes grading (slopes) but it includes also the curves and the creation of tunnels or other features along the highway route. The geographic restrictions and layout of the area before attempting to determine the best route is necessary to be understood.

Due to the natural landscape and also existing buildings and other features, the few highways are able to run perfectly straight along their entire length. Among many jobs of surveyor's one of them during highway construction is to aid in the mapping of this route. Thus, the most challenging part of the highway planning is often the necessity of bridges or tunnels. The surveyors provide input on their location as well as monitor the highway construction process in order to ensure that they are located correctly.

After the dirt has been graded and leveled, a surveyor would again survey the area for ensuring that the land has been correctly prepared according to the plans. Thus, construction begins. But before the advent of GPS technology, it involved the use of surveying stakes to mark the road edges. Nowadays, GPS has eliminated this process, by making it much easier in order to ensure that the road is aligned with the plans.

After the road is built, the surveys can be required for a wide variety of reasons, including road repair or maintenance. The highway surveys are also used during the process of accident reconstruction and investigations.

Therefore the process of taking road measurements for constructing a survey once required closing of the entire road because the surveying team could set up anequipment on the road's surface. However today, the specialized laser equipment allows this to be done from the roadside, virtually by eliminating the need for lane closures because of surveying. Then the highly accurate data is delivered by these systems which allows for the creation of detailed 3D maps showing not only the roadway but also overhead wires, barriers and other features.

Highway Project Formulation

A new highway project comprises of design and construction either at network of new roads, or a road link. There may also be projects which require redesign and re-alignment of existing roads of upgrading the geometric design standards. A new highway project formulation may be divided into the following stages:

1. Selection of route and survey of possible alignment of roads.
2. Finalisation of highway alignment.
3. Collection of materials and testing of subgrade soil and other construction materials.
4. Design aspects.
5. Construction stages.

Selection of Route and Surveys

Based on the points to be connected, different routes are tried. Selection of route, the requirement of alignment and the geological, typographical and other features of the locality are kept in mind. Surveys to be conducted for selection of route comprises of the following steps:

1. Map Study

This is the preliminary stage of the project.

At this stage the help of available maps of the area are studied.

2. Reconnaissance Survey

Tentative alignments are primarily looked into keeping in various factors. To decide the correct alignment a reconnaissance survey is conducted.

Following are the points to be considered during the reconnaissance survey:

1. The alignment should be such that necessary cutting and banking are avoided.
2. Obstructions to be faced along the alignment are looked into.
3. Gradients and length of curves to be provided in the alignment.
4. The alignment should avoid any cross-drainage work.
5. The alignment should not be taken through agricultural land completely.
6. The type of soil along the route should be looked into.
7. The sources of construction materials along the route to be assessed.
8. The type of terrain to be noted.
9. The magnetic bearing of the alignment are measured by compass.
10. Nature of ground on either side of the alignment and the highest flood level are noted.

3. Preliminary Survey

The preliminary survey or the preliminary location survey of the highway project is carried out with the following main objectives:

1. Various alternate alignments.
2. Estimation of the quantity of earthwork materials and other construction aspects.
3. Compare different proposals.

Following surveys are conducted during the preliminary survey:

1. Primary traverse.
2. Topographical surveys.
3. Levelling work.
4. Collecting hydrological data.
5. Soil surveys.
6. Establishment of permanent benchmarks by running a fly levelling.
7. Intersection of canals, of rivers are to be properly located.
8. Drawings are prepared following route survey, longitudinal and cross- sections, mass hall diagram, contour map of the strip and design of curves.
9. At the office, the cost estimate of different works such as land compensation, flyovers, culverts, bridges, earthwork, etc., are prepared.

1.5.1 Highway Alignment

Finalization of Highway Alignment

After the preliminary surveys, the final alignment is decided and is first located on the field by establishing the center line. Next the detailed survey should be carried out and the geometric design details are made. The detailed survey involves the following:

1. Centre line is marked at intervals of 30 m.
2. Fixing temporary benchmarks along the route for every 300 m.
3. Cross-section details are taken for 30 m on either side of the central line.
4. All details of cross drainage works are taken.
5. Topographical details are taken.
6. Detailed soil survey is carried out.
7. Total land width required is marked by pillars at regular intervals.
8. Tangent points and intersection points of the curves are properly marked by pillars.

Collection of Materials and Testing

Soil samples collected during the survey along the selected route are tested in the laboratory in order to design the pavement and the design of embankments and cuttings. Selected construction materials from borrow areas are also tested for the required properties.

Mix-design tests of materials are conducted in the laboratory to explore the possibility of using them for the pavement component layers. The possibility of using low-cost construction material like soil aggregate mixes, soft aggregates, stabilized soil and concrete mixes, in the sub-base or base course layers of pavement should also be fully explored.

1.5.2 Factors Affecting Alignment

An alignment should be straight and short. Though not practically feasible due to various practical difficulties it has to be satisfied along the route. A short route may have a very steep gradient which cannot be accepted as the vehicles have to spend more fuel and at the same time the vehicle may undergo more strain.

Short route may face some construction and maintenance problems. Shortest routes are deviated to satisfy certain obligatory factors. Sometimes long straight routes on a flat terrain are monotonous to drivers. Hence after a few kilometers it is desirable to have a slight bend to break the monotony and keep the driver alert.

A road which is economical now need not continue to be economical in maintenance or the vehicle operations cost. A straight and short alignment may not be economical due to the ground and weather conditions.

Hence an alignment may not satisfy all the requirements and factors but has to exercise a proper judgement to decide considering all the factors. The following factors such as technical, obligatory, economy and other considerations control the highway alignment.

They are:

1. Traffic
2. Geometric Design
3. Obligatory factors
4. Economy
5. Other considerations

Traffic

It is the prime factor to be considered. It comprises of type of vehicle it has to cater and the volume of traffic it has to handle. Volume of traffic is an important factor that specifies the quantity of movement per unit of time at a specified location.

Origin of the volume of traffic and destination study should be carried out. On the stipulated area, the desired lines be considered along with the traffic flow. Proposed new road alignment should also take into account the traffic flow patterns.

Future plan as per the National road plan should also be considered e.g. National Highway, Major District Highway etc.

Geometric Design

It is the other technical factor which should be given due considerations. Geometric design factors such as gradient, radius of curvature, formation width, carriageway, right of way, sight distance, etc. would be considered in the alignment of the road.

In general the gradient to be less than the ruling gradient or design gradient for a smooth movement of vehicles. Extra widening of the road along the curve, the superelevation to be provided, speed of vehicle, etc. are to be studied after fixing an alignment.

If the alignment is not satisfactory for the above geometric design factors, the alignment may have to be changed. Studies are to be made at horizontal curve locations and vertical curve locations to find whether the alignment will be satisfactory or not.

Line of sight, sight distance and stopping sight distance which will be available at every section should be studied. Overtaking operations of vehicles should be smooth for the permissible speeds of the vehicles on the road at the selected alignment.

Obligatory Factors

Apart from the general requirements certain requirements are mandatory or control points for the alignment of a highway.

They may be divided broadly into two categories viz.,

1. Points through which the alignment should pass.
2. Points through which the alignment should not pass.

1. Obligatory Positive Factors

In order to satisfy the obligatory points through which the road alignment should pass,it may cause the alignment to deviate from the shortest or the economical path.

Following are some of the obligatory points through which or near to the alignment should pass.

(i) Hill or Ridge

When the alignment has to cross a hill or a ridge, the alternatives are:

i. To cut a tunnel.

ii. To go round the hill.

iii. To deviate fully till a suitable hill pass is available.

Figure shows the deviation of alignment AB to ACB and thereby avoiding a tunnel or large cutting.

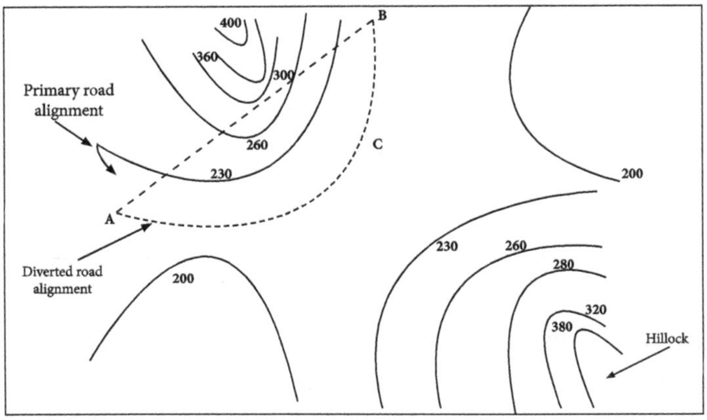

Alignment along a hill side.

(ii) River Crossing

Here the road has to cross in general perpendicular to the river center line at that point. Only at the location,bridge can be constructed. If such a point is not available oblique bridge alignment is made. For sight distance and line of sight considerations, perpendicular straight intersections are preferred.

The bridge location should have good foundation so as to be economical. Hence a straight alignment AB has to be taken along ACB. Alterations are also needed in case of railway, irrigation and navigation canals.

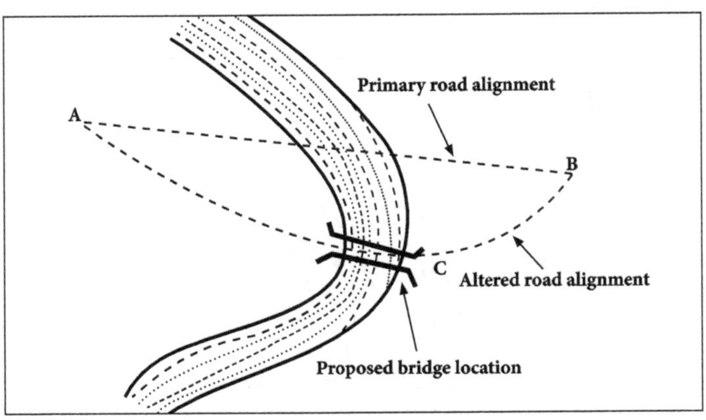

Alignment to suit proper location of bridge.

(iii) Connecting to Stations

The main aim of road alignment is to have short path, sometimes it may be necessary to connect some of the towns and villages to have better service of the road. In such a case, the road has to be diverted to suit the situation. Thus the straight alignment AB is diverted through a town C which is being linked to two villages D and E.

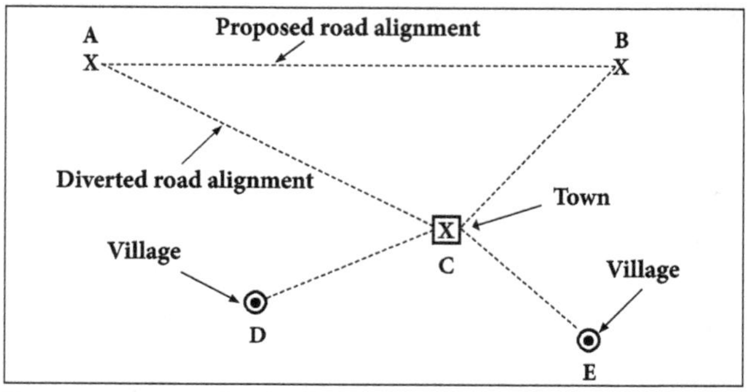

Alignment connecting intermediate towns and linking villages.

Such deviations are to be made to give a spectacular view of a canyon for a scenic park and connect an agricultural land or a farm road.

2. Obligatory Negative Factors

Conditions may also arise in practice that make to deviate from short route. Such situations are discussed below:

i. Religious places

ii. Monuments

iii. Unsuitable land

iv. Large water body

v. Agricultural land

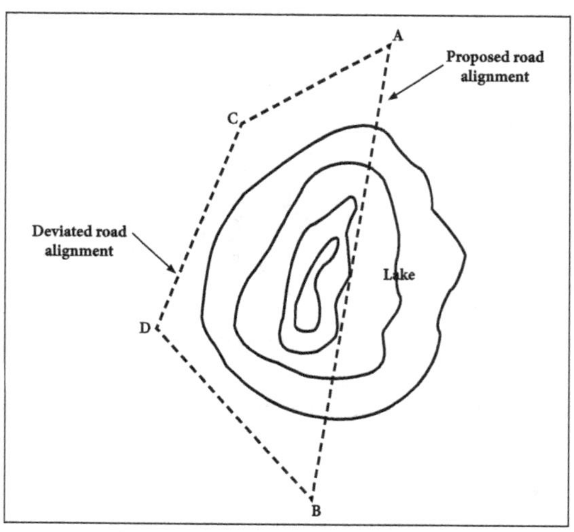

Alignment deviating from a lake.

Economy

The two technical and obligatory factors considered above should be basically economical. While working out the economic aspects of the road alignment, the initial cost, the maintenance cost and vehicle operations cost should be taken into account. Initial cost of construction may be reduced by slightly altering the alignment such that construction of embankment or heavy cutting may be avoided or atleast reduced. A balanced alignment may be made by equalising the cost of cutting and construction of embankment.

Other Considerations

Factors like drainage considerations, hydrological factors, monotony, etc. also govern the final alignment. Drainage considerations, sub soil conditions, cross-drainage works, crossings may also affect the alignment. Border roads may pose special problems according to the relations with the neighboring countries.

1.6 Engineering Surveys

Conventional Engineering Surveys for Alignment

Engineering surveys are to be carried out before finalizing any highway alignment. The first step in highway alignment and design is to procure all available pertinent data. Engineering surveys for alignment comprise of four stages. The four stages are:

1. Map study
2. Reconnaissance survey
3. Preliminary survey
4. Final location survey

Of the above four stages, first three stages of study and survey are conducted to arrive at alternate alignments keeping in view the salient features. The fourth stage of survey goes into conducting a detailed survey on the selected alignment. Various aspects to be considered in each stage are as follows:

Map Study

Different form of maps, aerial photographs, charts or graphs, etc are collected. Topographical maps prepared by Survey of India, Government of India, are available with 15 or 30 meter contour intervals. They also provide the details of rivers, hills, valleys.

Careful study of the maps at the office, gives the basic details to fix serious alternate alignments. Details of these alignments may be obtained on the site. Possible alignments can be decided based on the following details available on the map. The details are:

1. Alignment bypassing valleys, lakes or ponds.
2. Availability of a mountain pass in a mountain range.
3. Deciding the location of bridges for crossing rivers avoiding de-touring of road.
4. Connecting two station points at different elevations by adopting a suitable permissible or ruling gradient.
5. Computation of length of roadway using the contour interval and the ruling gradient.

Basic study of topographic maps enables to select alternate routes which are feasible to proceed. This basic selection of route should be further studied in the field.

Reconnaissance Survey

Reconnaissance survey may be aerial reconnaissance or ground reconnaissance.

Aerial Reconnaissance: Provides a bird's eye view of the alignments under consideration. It also gives the details of surroundings along the routes. Detailed study of the reconnaissance may enable for change of alignments based on the actual features seen. A final selection may be made by a close view of the selected alignments.

Ground Reconnaissance: It consists of general examination of the ground by walking or riding along the selected routes.

The fixed survey party inspect in parallel different routes based on the alignments proposed on the basis of map study.

Each group collects all available information necessary for evaluating the merits and demerits of each proposed alignment. In order to identify the reference points pegs should be driven at different locations.

Simple instruments like Abney level, Tangent clinometer, Pedometer, Aneroid barometer, Ghat tracer, etc are carried by the survey party. The details to be collected during ground reconnaissance are given below:

(i) Topography: It includes plain, rolling, steep valleys, ponds, lakes, marshy land, cliffs, gorges, etc.

(ii) Geometries: It includes gradients, radius of curves, hair-pin bends, right of ways, constraints like monuments.

(iii) Roads: It includes types of roads such as jeep track, earthen track, length along each alignment, existing conditions, elevations, etc.

(iv) Soil Type: It includes nature of soil, marshy land, flooding area, sand dunes, black-cotton soils, field identification, geological features, etc.

(v) Climatic Conditions: It includes monthly maximum and minimum temperatures, rainfall, average annual peak period, wind direction and velocities, fog conditions, sunny days, snow fall, etc.

(vi) Crossings: It includes cross-drainage structures, railway crossings, bridge and culverts, maximum flood level, ground water level, water table fluctuation, existing highway crossings, etc.

(vii) Resource: It includes construction materials, labour-availability, locations of stone quarries availability of water, communication facilities, etc.,

(viii) Economic Factors: It includes population, marketing centre, land value, agricultural and economic potential of area, environmental factors, etc.,

(ix) Obligatory Points: Ancient monuments, burial grounds, religious structures, hospitals, schools, location of towns and villages, agricultural lands, industrial areas, etc., Ground reconnaissance may be followed by an aerial reconnaissance if the area is vast. The alignments proposed based on the map study may be altered on the basis of the reconnaissance. A few of the alternate alignments may be chosen for further study based on the details available from the reconnaissance survey.

Preliminary Survey

Preliminary survey is the important survey which decides the final location.

This is conducted with the following objectives:

1. The proposed alignments based on reconnaissance survey are accepted at this stage and all the necessary details of topography, soil and drainage details are collected.
2. The proposed alignments with the background details are checked with general requirements needed for an ideal alignment.
3. The details of construction cost of every proposed alignment is made.
4. The best alignment from all considerations is finalized.

This survey is conducted on the finalized alignment with large scale instruments.

It may be carried by following conventional approach or modern approach depending on the magnitude of the project. The conventional approach has all the following components, viz., primary traverse, topographical features, levelling work, hydrological data and Soil survey. Modern rapid approach is by conducting aerial survey by adopting aerial photographs.

Conventional Preliminary Survey

It is discussed in the following paragraphs:

(i) Primary Traverse

An open traverse is conducted along the alignments. The length of center line should be measured by using chaining methods more accurately. Instead a tachometric survey may be conducted. For alternate alignments a secondary traverse or an independent primary traverse may be necessary.

(ii) Topographical Features

In the next set up, the topographical features are recorded soon after establishing the center lines by the primary traverse.

Details along the center line such as geographical and other manmade features on either side of the central line are surveyed and plotted.

Width of survey to be conducted is the minimum land width or as decided by the surveying party.

(iii) Levelling Work

Levelling work is also to be carried out to get the central line profiles and typical cross sections. A minimum levelling work is done along the alternate alignments to obtain the approximate earthwork.

(iv) Hydrological Data

The next step is to decide the type, number and approximate size of cross drainage structures. For this necessary hydrological data are collected. For the vertical alignment, the grade line is decided considering the hydrological and drainage data.

(v) Soil Survey

Decision of the alignment depends on the type of soil on which the maximum length of road alignment passes.

This survey conducted at this stage is helpful for the following activities:

- To work out the details of earthwork.
- To find the suitability of materials for construction.
- To check the slope, gradients and their stability.

- To decide the type of pavement.
- To decide the base and sub-base courses.
- To find the surface drainage requirements etc.
- To get the soil details, a detailed soil sampling need not be done at this stage but auger boring may be enough.

Samples all along the alternate alignments may be taken to get the field identification followed by laboratory basic tests to get preliminary idea of the soil profile. Geophysical and electrical resistivity methods may yield the location of hard stratum and thickness of other layers. Soil samples collected on the soil survey are used to get the soil consistency and classification. Soil profile is drawn roughly all along the alternate alignments up to the depth of exploration.

Aerial Preliminary Survey

This is conducted if the area to be surveyed is vast by following the steps given below:

1. Photographs of the strip of land is taken with the required longitudinal and lateral overlaps. In order to prepare mosaic,vertical photographs are needed.
2. The photographs are examined under stereoscopes and control points are selected. These are used for establishing the traverse of the alternate alignments proposed. The control points are located on the maps.
3. Spot levels and subsequently contour lines may be obtained using stereo-pair observations. Also the topographic details may be noted down on the maps.
4. Photo-interpretation techniques are used to decide the geological features, soil conditions, drainage requirements, etc.

With the data collected from the preliminary survey of all the alternate alignments, a comparison is made. The final selection of alignment is made keeping in view the requirements and the factors to be satisfied for the best selection of final alignment. Final location and detailed survey is conducted on this selected alignment.

Final Location Survey

1. Design Office Work

After the conclusion of preliminary survey, plans and longitudinal sections are prepared so as to go ahead for detailed location survey. Sharp curves, hair-pin bends, bridge crossings, etc. are shown on the plan. Contours at 1 to 3 m intervals are drawn to facilitate the final decision.

Scales for the map be adopted as 1:1000 horizontal and 1:100 vertical. As far as possible the same scale is adopted in the final drawing also. On certain locations, e.g., at hair pin bends, steep terrain, etc. plans are required to be drawn on a smaller scale.

In general, for built up areas and hilly terrain a horizontal scale of 1 in 2500 for horizontal scale and 1 in 250 for vertical scale may be adopted.

2. Field Work

Based on the center line selected in the design office, the final survey is carried out to lay out the final center line of the road in the field.

During this survey necessary additional data for the working drawings are made.

If necessary minor modifications in the final location may be made.

Two operations are involved in this survey, viz.,

i. Staking out the final center line of the road by means of a continuous transit surveys.

ii. Detailed levelling.

3. Final Centre-line Marking

The decided central line is transferred on the ground by means of a continuous transit survey. As the survey proceeds, staking or pegging of the center line is done. All angles should be measured with a transit. All reference points should be securely located so as not to be disturbed during construction. These reference points are to be transferred on the final plan drawing.

Pegs should be fixed at 50m intervals in plain and rolling terrain and 10 to 20m intervals in hilly terrain. Distances measured along the final line should be continuous following the horizontal curves. At rail and road crossings, the angles subtended by the roads at the center line are measured with the help of a transit. Temporary bench marks are established at suitable intervals (about 300 m) to have vertical controls along the center line.

4. Detailed Survey

Levelling works play an important role as the vertical alignment, earth work computations, drainage design details are to be worked out.

Along the final center line, levels should be taken at all staked stations and at all locations where there are undulations in the ground cross-sections.

The following detailed survey are carried out:

i. Cross-section and profile.

ii. Surface drainage area.

iii. Soil suitability analysis.

iv. Coarse aggregates availability.

v. Right of way investigation.

vi. Investigation on crossings.

Modern Methods of Surveying

Photogrammetry, often called "Photographic remote sensing" is the science or art of obtaining measurements, needed for highway surveys, by means of photography. It is intended to encompass procedures for photo interpretation and for converting single photographic into composite ones called orthophotographs or mosaics and into maps.

It is based on aerial photographs. It is a basic working tool for the highway engineer. Application of remote sensing is not only helpful in location survey but also in planning, geometric design, rights of way, traffic studies, drainage, soil classification and identification, earthwork measurement,materials location and pavement condition surveys.

Vertical aerial photographs taken with the camera pointed nearly straight down are the most useful for highway mapping purposes. The area to be covered is photographed in parallel with the individual pictures lapped both in the direction of flight (end-lap) and between successive runs (side-lap).

For stereoscopic uses, end-lap must be greater than half the picture width in order that the center of one photograph is included in both adjacent photographs. Side-lap should be around 25% and percentages less than 15% more than 35% are unacceptable. Overlap can be stated in terms of acceptable base to height and width to height ratios. Height from which photographs is to be taken depend on the maps to be made.

The factors to be considered for the purpose of preparing maps are:

i. Local length of aerial camera.

ii. Desired combination of map scale.

iii. Contour intervals.

iv. Ratio of map scale to photograph scale.

Several instruments are available for converting data from aerial photographs to maps. They are multiplex, stereoscopic plotters and autograph. It is possible with any of these instruments to produce an accurate map showing all natural and artificial features. Also, contours may be drawn or spot elevations be determined.

1.6.1 Drawings and Reports

Design Aspects

After the survey and investigation are over, a new project involves in the following design steps:

1. Identifying the class of road to be designed and the terrain over which it has to pass.
2. Design speed selection appropriate to the class of road and the terrain.
3. Selection of cross-sectional elements keeping in view the future traffic requirements.
4. Selection of appropriate geometric design standard suitable to the class of road, terrain and design speed.
5. Design of alignment and profile.
6. Alignment design of embankments and deep cuts.
7. Choosing of appropriate pavement based on the need and material availability.
8. Design of pavement.
9. Design of surface and subsurface drainage system and erosion control.
10. Design of cross-drainage structures.
11. Design of road junctions.
12. Estimation of various quantities of items and the total cost.
13. Estimation of road use cost.
14. Estimation of maintenance cost.
15. Comparing various alternative and selecting the optimal solution.
16. For the selected route, preparation of final drawings and estimates.
17. Preparation of final project report.

Construction Stages

The construction of the road may be divided into two stages, viz.,

i. Earth work.

ii. Pavement construction.

The earthwork comprises of excavation and construction of embankments.

During the excavation process for highway cuts, protection of earth slopes and construction of drainage works are taken care of and construction of highway embankment by rolled-fill method is preferred. Compaction of soil in layers under controlled moisture and density are done using suitable layers. In situations of high embankments special checks such as stability of the embankment foundation and slopes, settlement, etc. are to be done.

2

Highway Geometric Design

2.1 Importance of Geometric Design: Design Controls and Criteria

The geometric design of the highways deals with the dimensions and layout of visible features of the highway. The emphasis of the geometric design is to look into the requirement of the driver and the vehicle such as safety, comfort, efficiency, etc.

The features generally considered are the cross section elements, sight distance consideration, horizontal curvature, intersection and gradients. The design of these features is to a great extend influenced by driver behavior and psychology, vehicle characteristics, traffic characteristics like speed and volume.

Proper geometric design will help in the reduction of accidents and their severity. The objective of geometric design is to provide optimum efficiency in traffic operation and maximum safety at a reasonable cost. The planning cannot be done stage wise in this case like that of a pavement, but it has to be done well in advance.

Factors Affecting Geometric Design

A number of factors affect the geometric design are as follows:

Design Speed

Design speed is the single most important factor that affects the geometric design. It directly affects the horizontal curve, sight distance, and the length of vertical curves. As the speed of vehicles vary with terrain, driver, etc, a design speed is adopted for all the geometric design.

Design speed is defined as the highest continuous speed at which an individual vehicles can travel with safety on the highway when the weather conditions are conducive. The design speed is different from the legal speed limit which is the speed limit imposed to curb a common tendency of the drivers to travel beyond an accepted safe speed. Design speed is also different from the desired speed which is the maximum speed at which a driver may travel when unconstrained by either traffic or local geometry.

Since there are wide variations in the speed adopted by different drivers, and by different types of vehicles, design speed must be selected such that it satisfy nearly all drivers. At the same time, a higher design speed has cascading effect in the other geometric designs and thereby cost escalation. Therefore, an 85th percentile design speed is normally adopted. This speed is defined as the speed which is greater than the speed of 85% of drivers. In some countries this is as high as 95 to 98 percentile speed.

Topography

Another important factor that affects the geometric design is the topography. It is easier to construct roads with the required standards for a plain terrain. However, for a given design speed, construction cost increases multiform with the gradient and the terrain. Therefore, geometric design standards are different for different terrain to keep cost of the construction and time of construction under control. This is characterized by sharper curves and steeper gradients.

Other Factors

In addition to design speed and topography, there are various other factors that affect the geometric design and are as follows:

Vehicle: The dimensions, weight of the axle and operating characteristics of a vehicle influence the design aspects such as width of the pavement, clearances, parking geometrics, radii of the curve, etc. A design vehicle which has standard weight, dimensions and operating characteristics are used to establish the highway design controls to accommodate vehicles of a designated type.

Human: The important human factors that influence the geometric design are the physical, psychological and mental characteristics of the driver and pedestrians like the reaction time.

Traffic: It will be uneconomical to design the road for peak traffic flow. Therefore a reasonable value of the traffic volume is selected as the design hourly volume which is determined from the various traffic data collected. The geometric design is thus based on this design volume, capacity etc.

Environmental: Factors such as air pollution, noise pollution etc. should be given due consideration in the geometric design of roads.

Economy: The design adopted must be economical as far as possible. It should match with the funds alloted for capital cost and maintenance cost.

Others: Geometric design should be such that the aesthetics of the region is not affected.

2.2 Highway Cross Section Elements

Developing a roadway design focuses on the selection and configuration of the elements that comprise the roadway cross section. Once defined these elements are collectively referred to as the typical sections. Cross-section elements directly relate to the number of lanes to be provided, the width of pavement, location of shoulders or curbs, medians, slopes, cambers, embankment, drains, etc. Each of the elements are discussed in the following paragraphs.

Right of Way

Right of way is the width of land acquired permanently to be designated as Road Land. Road land requirement depends on the future development of the highway based on the traffic load.

The factors that have influence in taking a decision are:

1. Width of formation
2. Shoulders
3. Class of road
4. Expected traffic
5. Availability of land
6. Availability of funds

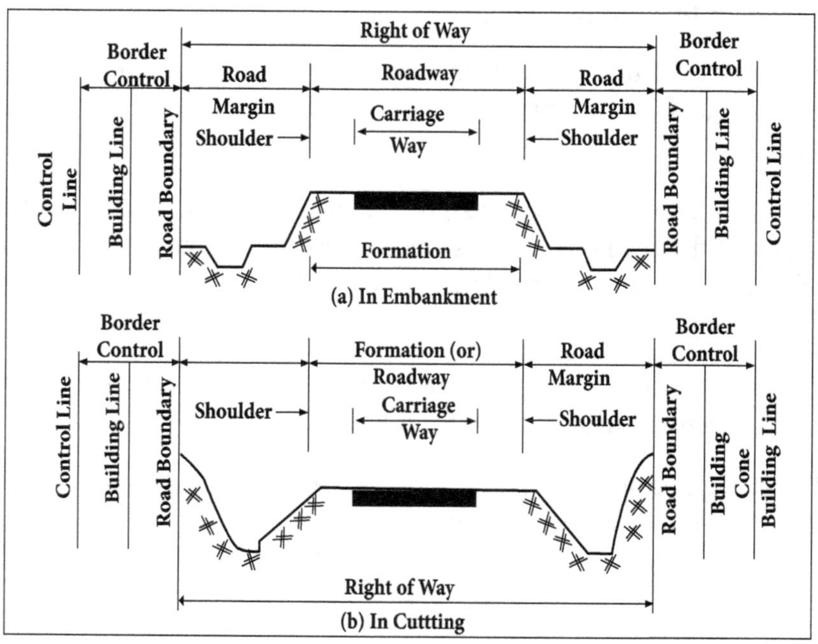

Right of way: (a) In embankment and (b) In cutting.

These additional strip of land is required to dump the surplus excavated earth as spoil bank or for obtaining extra earth for road work. Figure shows the position of right of way in embankment and in cutting. Table shows the requirement of right of way for different types of highways on various terrains as per IRC specifications.

Table: Requirement of Right of Way.

Road Classification	Plain and rolling terrain m				Mountain and steep terrain m	
	Open areas		Built-up areas		Open areas	Built-up areas
	Normal	Range	Normal	Range	Normal	Normal
National and State Highways	45	30-60	30	30-60	24	20
Major District Roads	25	25-30	20	15-25	18	15
Other District Roads	15	15-25	15	15-20	15	12
Village Roads	12	12-18	10	10-15	9	9

Border Controls

In order to reserve sufficient space for future improvement of roads, the building construction activities have to be controlled on either side. Building activities should not be allowed beyond building line.

Table: Recommended Border Controls.

Road Classification	Plain and rolling terrain, m			Mountain and steep terrain, m	
	Open areas		Built-up areas	Distance between building line and road boundary in, m	
	Overall width between building lines, m	Overall width between control lines, m	Distance between building line and road boundary, in m	Open areas	Built-up areas
National and State Highways	80	150	3 to 6	3 to 5	3 to 5
Major District Roads	50	100	3 to 5	3 to 5	3 to 5
Other District Roads	25/30	35	3 to 5	3 to 5	3 to 5
Village Roads	25	30	3 to 5	3 to 5	3 to 5

If the land width is equal to the width between building lines indicated in this column, the building lines should be set-back 2.5 m from the road land building.

The overall requirement of width as per IRC standards are presented in table. The normal land width required for the National and State Highways on open plain terrain is 45 m and the maximum land width is 60 m. The corresponding width between lines and control lines are respectively 80 m and 150 m. The set back allowed is 10 m and 45 m beyond the road boundary lines keeping the maximum road width.

Carriageway

The pavement width or carriage width depends on the width of traffic lane and number of lanes. Carriageway accommodating one line of traffic movement is called as a traffic lane. Width of the lane is decided based on the width of vehicle and the minimum side clearance which is provided for safety considerations.

Increase of side clearance may allow more vehicles with high speed. Keeping these aspects in view a carriageway width of 3.8 m is considered desirable for a single lane for vehicle of maximum width 2.44 m.

For two or more lanes,the carriage width may be 3.5 m per lane. As per IRC specifications, the maximum width of vehicle is 2.44 m. For a single lane, the carriage width is 3.8 m and the side clearance is 0.68 m. For a two lane, 7.0 m is provided as carriageway and 0.53 m as side clearance and 1.06 m in between vehicles. The width of carriageway for different lanes or roads are given in table as per IRC standards.

Table: Recommended width of carriageway.

Class of Road	Width of Carriageway
Single lane	3.75 m for all roads; may be decreased to 3 m on village roads
Two lanes without raised kerbs	7.0 m
Two lanes with raised kerbs	7.5 m
Intermediate carriageway (except on important roads)	5.5 m

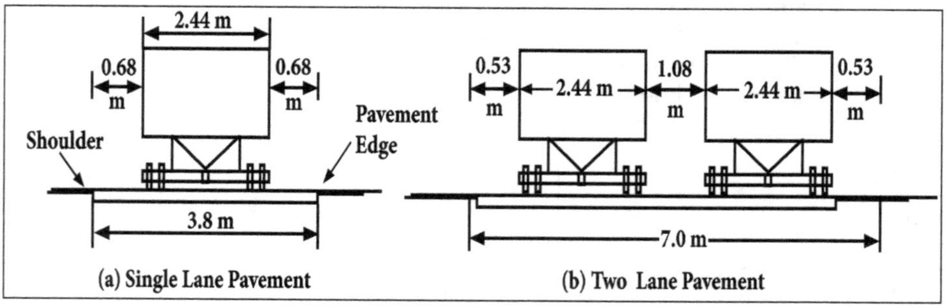

Lateral placement of vehicles.

In some highways traffic separators or meridian are provided. In such cases the road width depends on the pavement width (or the lane width and number of lanes) and the width of traffic separators. The width of pavement is increased on horizontal curves.

Kerbs

Kerbs indicate the boundary between the pavement and shoulders. Kerbs are provided on islands or footpaths. Based on their functions, kerbs may be divided into three groups as given below:

1. Class I or Low or Mountable Kerb

It helps the driver to enter the shoulder area with little difficulty. The height of this class of kerbs is 7 to 8 cm. Kerbs are also useful for longitudinal drainage system.

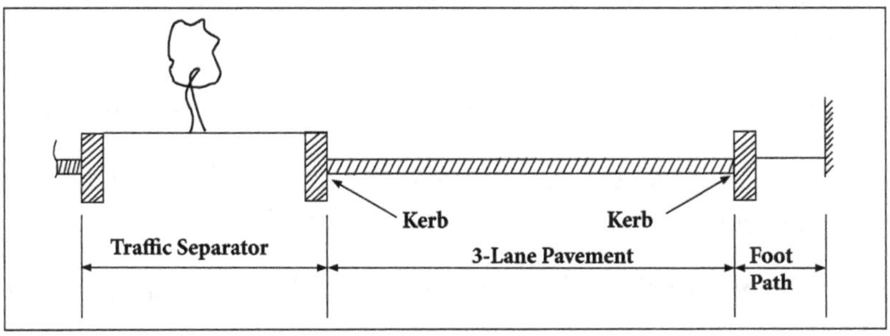

Kerb and traffic separator.

2. Class II or Low Speed Barrier or Urban Parking Kerb

It consists of a vertical face of 15 to 20 cm high with 2.5 cm batter. This shape of arrangement is made in the kerb to prevent scraping of tyres.

This type prevents encroachment of slow speed vehicles or parking vehicles to the foot path or parking space. In emergency conditions, the driver can drive over through this curb.

3. Class III or High Speed Barrier

It is considerably high with height of 23 to 45 cm or more so as to prevent vehicles leaving the roadway under any circumstances. It is particularly used in critical locations such as bridges or hill roads. Kerbs are provided to offer lateral stability to granular base courses and flexible pavements. Submerged kerbs are sometimes provided at pavement edges between the pavement edge and shoulders.

Shoulders

Road margin is the term collectively used for the facilities provided on either side of a road pavement, viz., shoulders, parking lane, frontage road, driveway, cycle track, foot path, guard rail and embankment slopes.

Shoulder is that portion of the roadway which lies between the outer edge of the carriageway and the inner edge of the kerb, ditch gutter or slope.

Shoulders are made out of earth which are of very narrow width of 1.2 m. It can be used only in fair weather. Width recommended by IRC is 2.5 m. Larger shoulder widths are now being recommended.

Shoulders are basically provided to serve as an emegency lane for vehicles compelled to be taken out of the pavement and roadway and also serve for vehicles which have broken down.

Factors to be considered for Shoulders:

1. It should have adequate bearing capacity so as to support loaded trucks even in wet weather.
2. Its use increases rapidly with a decrease in pavement width below 7 m.
3. Well maintained grass shoulders have the same effect as that of gravel shoulders with respect to the effect on transverse position.
4. Surface of shoulders should be rougher than the traffic lane so as to discourage the drivers.
5. To distinctly identify, the colour of the shoulder should be different from that of the pavement.
6. Bituminous treated shoulders with width more than 1.2 m adjacent to a 7.0 m pavement serve to increase the effective width by 0.6 m.
7. The width of shoulder should be around one-half the difference between the roadway width and carriageway width.

Foot Paths

Foot paths also called as side walks are provided in roads constructed in urban areas when the vehicular and pedestrian traffic are heavy. It is provided to give adequate protection to pedestrians and thereby decrease accidents. Side walks are provided on either side of a road and the minimum width is 1.3 m. Footpaths should be very attractive such that the pedestrians will prefer to use the foot path than the roads.

Separators

Separators are provided to prevent the head on collision between two vehicles moving in opposite directions in the adjacent lanes.

Separators may be in the form of pavement markings, physical dividers or area separators.

Out of these three, pavement marking is the cheapest. Physical separators like mechanical separators should be designed in such a way that the vehicles should not be damaged. Wide area separators of 8 to 14 m width may be considered depending on the availability of land.

To reduce the headlight glare, a minimum width of 6 m is required. Figure shows kerb and traffic separator. Indian Roads Congress recommends a minimum desirable width of 5.0 m for rural highways.

If adequate land is not available this may be restricted to 3.0 m. On long bridges the separators are reduced to 1.2 to 1.5 m. The separators should be of uniform width on a particular road.

Camber

Also called as cross-fall, is the convexity provided to the cross-section of the surface of carriageway. It is the difference in level between the highest point, known as the crown usually located at the center of the carriage and the edge.

Camber is provided so as to perform the following activities:

1. To drain surface water.
2. To separate the traffic in two opposite directions.
3. To improve the appearance of the road.

First reason to provide a camber is very much important and this will require a profile which can remove water quickly to the gutters. Amount of camber to be provided depends on the intensity of rainfalls, rate of run-off and type of road structure.

Second reason provides a natural separating point for the traffic. Third reason adds to the better appearance of cross-section. A steep camber is to be preferred in areas of heavy rainfall and for a poor surface material used for the road.

For a better surface material used for the road instruction, such as a concrete road, a flat camber may be enough. Table below shows the minimum camber requirement as per Indian Roads Congress for different types of highway surfaces.

Table: Camber/Cross-fall for different road surfaces.

Surface Type	Camber / Cross-fall
Cement concrete or High type bituminous surface	1.7 to 2.0 % (1 in 60 to 1 in 50)
Bituminous surface	2.0 to 2.5 % (1 in 50 to 1 in 40)
Water Bound Macadam, Gravel surface	2.5 to 3.0 % (1 in 40 to 1 in 33)

Earth surface	3.0 to 4.0 % (1 in 33 to 1 in 25)

Different types of cambers are shown in figure.

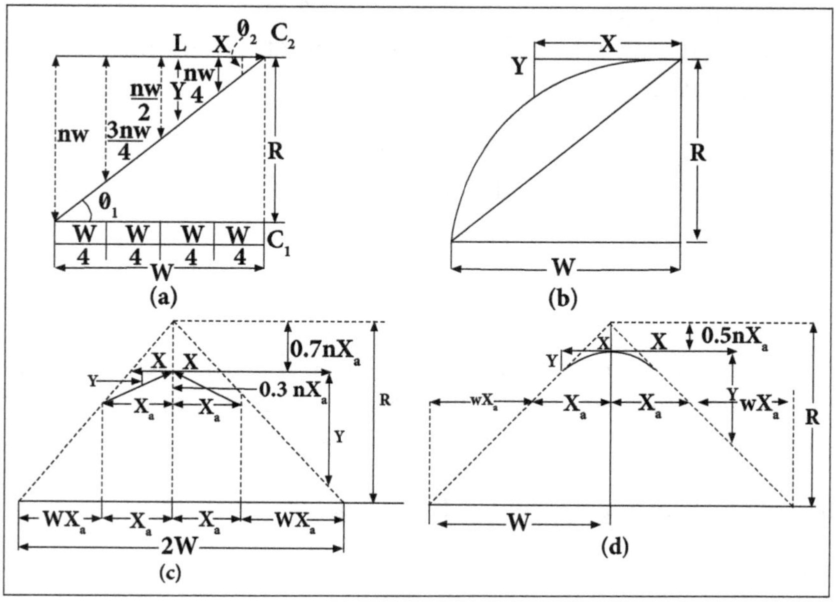

Types of cambers.

2.3 Sight Distance Elements

Visibility is an important requirement for the safety of travel on highways. For fast moving vehicles and travel during night times, the visibility is of paramount importance.

Visible road length should be sufficient enough such that the driver of a vehicle will be in a position to see any obstruction on the road without any difficulty. There should be adequate length of road visible ahead to the driver at any distance. This length of road is called as sight distance.

Sight distance is the actual distance available from a point along the road surface at which a driver from a specified height from the road, has visibility of a moving or stationary object on the road. Examples of sight distance restriction on horizontal curves or on vertical curves or at road intersections are shown in figure.

Requirements to sight distance to drivers is needed for traffic control and for geometric design. In the geometric design, three sight distance situations considered are given below:

1. Stopping or absolute minimum sight distance.

2. Safe overtaking or passing sight distance.
3. Safe sight distance at uncontrolled intersection.

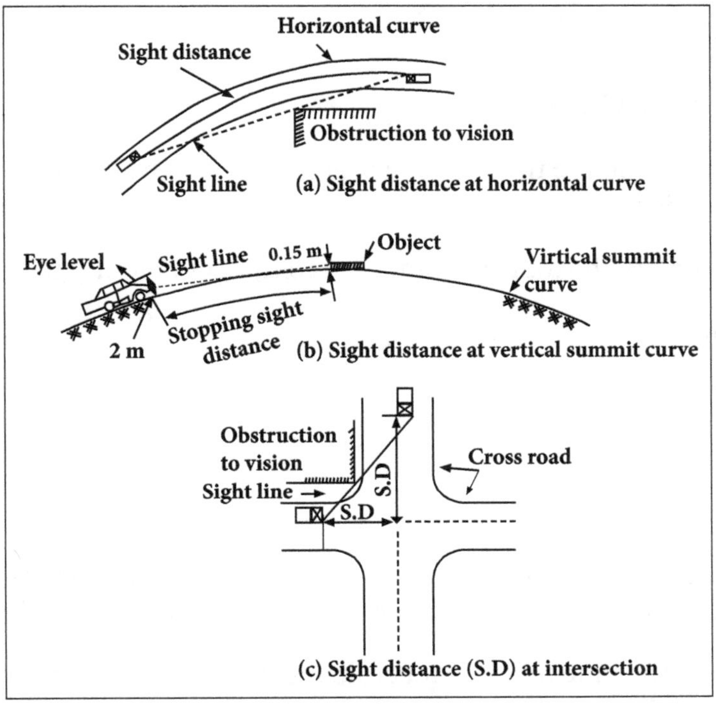

Sight distance restrictions.

2.3.1 Stopping Sight Distance

A driver on a vehicle driving at the design speed on a highway should have sufficient sight distance or length of road visible ahead to stop the vehicle in case of any obstruction on the road without collision. The stopping sight distance is the distance covered by a vehicle from the instance a danger is fully realized by the driver to the actual stop. This is also called as non-passing sight distance.

The sight distance available on a road to a driver depends on the following factors:

1. Features of the road ahead.
2. Height of the driver's eye above the road surface.
3. Height of the object above the road surface.

Stopping of a motor vehicle on a road after seeing an obstruction depends on the following factors:

1. Total reaction time of the driver.
2. Speed of vehicle.

3. Brake efficiency of the vehicle.
4. Frictional resistance between road and the tyres.
5. Gradient of the road.

1. Total Reaction Time

It is the time taken from the time the object is seen by the driver to the time the brakes are effectively applied.

As per the PIEV theory the total reaction time comprises of four parts, viz.,

i. Perception
ii. Intellection
iii. Emotion
iv. Volition

Perception time is the time required for transmitting the sensation received by the driver's eyes or ears to the brain through the nervous system and spinal chord.

Intellection time is the time required for realizing the situation. It is also the time taken for comparisons of different thoughts and alterations, regrouping and registering new sensations.

Emotion time is the time elapsed during emotional sensations and disturbance such as fear, anger or any other emotional feelings with reference to the situation. Volition time is the time taken for the final action. The PIEV process is illustrated in figure.

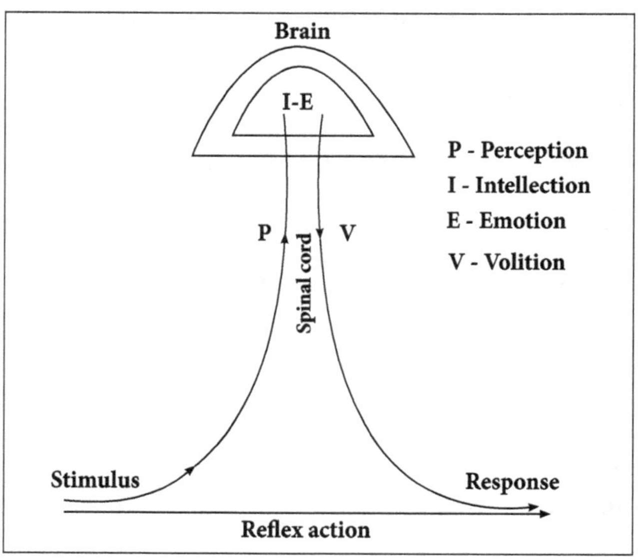

Reaction time and PIEV process.

The PIEV time of a driver depends on several factors such as mood, temperament and physical conditions of the driver, environmental factors, etc. For an average driver, the total reaction time varies from 0.5 seconds to 4 seconds depending on the complexity of the problem.

2. Speed of Vehicle

The distance moved by the vehicle during the reaction time is a function of the speed of the vehicle. After the application of the brake the distance moved by the vehicle before coming to a stop is based on the initial speed of the vehicle. Higher the speed more will be the stopping distance.

3. Brake Efficiency

If the brakes are in good condition, the wheels are fully locked once the brake is applied preventing the wheels from rotating. It will lead to 100% skidding which is not generally desirable and dangerous. In order to avoid skidding, the force due to brakes should not exceed the frictional force between the wheels and tyre.

4. Frictional Resistance between Road and Tyre

The coefficient of friction between the road surface and the tyres of a vehicle depends on the following:

i. The type and the condition of road surface.

ii. Presence of moisture, mud, snow, etc on the road surface.

iii. Conditions of treads of tyres.

iv. Tyre pressure.

Indian Roads Congress recommends a value of 0.40 at 20 km/hr to 0.35 at 100 km/hr for the frictional coefficient. Equating the work done by friction to the difference in the kinetic energies of the vehicle at the time of applying brakes and at the time of stopping, gives the braking distance. In general a rate of retardation of 3.93 m/sec^2 is considered comfortable to passengers.

5. Gradient of the Road

Depending on ascending or descending gradient,the gravity adds to or against braking action.

In ascending gradient,the gravity adds to the braking action and the breaking distance is decreased.

Components of Stopping Sight Distance

The total stopping distance of a vehicle consists of the following components:

1. Lag distance
2. Braking distance

1. Lag Distance: The lag or reaction distance is the distance moved by a vehicle during the total reaction time.

2. Braking Distance: Braking distance may be computed by equating the workdone in stopping the vehicle and the kinetic energy for a level road. Several factors contribute for the friction on the road. Let f be the coefficient of friction. The friction decreases with the increase in speed of the vehicle.

IRC has recommended values of friction for different design speeds table.

Table: Coefficient of Friction For Different Speeds.

Speed, km/hr	20 to 30	40	50	60	65	80	100
Longitudinal coefficient of friction, f	0.40	0.38	0.37	0.36	0.36	0.35	0.35

Table: SSD values for Different Speeds.

Design speed, kmph	20	25	30	40	50	60	65	80	100
Safe SSD for design, m	20	25	30	45	60	80	90	120	180

Problems

Design speed = 80 kmph, Total reaction time = 3 sec, Coefficient of friction = 0.6. With the following specifications, let us calculate SSD for:

(i) A two lane road with two-way traffic and

(ii) A single lane road with two-way traffic.

Solution:

Given:

Design speed = 80 kmph

Total reaction time = 3 sec

Coefficient of friction = 0.6

Formula to be used:

Assume the road is a level road,

Then, Stopping Distance = (Lag distance) + (Braking distance)

$$V = 80\,\text{kmph or } v = \frac{80}{3.6} = 22.22\,\text{m/sec}$$

$$t = 3\,\text{sec},\ g = 9.8\,g/cm^2 \text{ and } f = 0.6$$

Therefore, Stopping Distance $= vt + \dfrac{v^2}{2gf}$

$$= 22.22 \times 3 + \frac{(22.22)^2}{2 \times 9.8 \times 0.6}$$

$$= 66.66 + 41.98$$

$$= 108.64\,\text{m}$$

i. Stopping sight distance for two lane road with two way traffic = Stopping distance = 108.64 m.

ii. Stopping sight distance for single lane road with two way traffic = 2 × 108.64 = 217.28 m.

2.3.2 Overtaking Sight Distance and Intermediate Sight Distance

Overtaking sight distance is the minimum sight distance that should be available for a driver to overtake another vehicle safely on a two-way road.

It is also referred to as safe passing sight distance. Considering the driver's eye position as 1.2 m above road surface, the overtaking sight distance can be defined as the distance along the center line of the road with driver's eye level at a height of 1.2 m above the road.

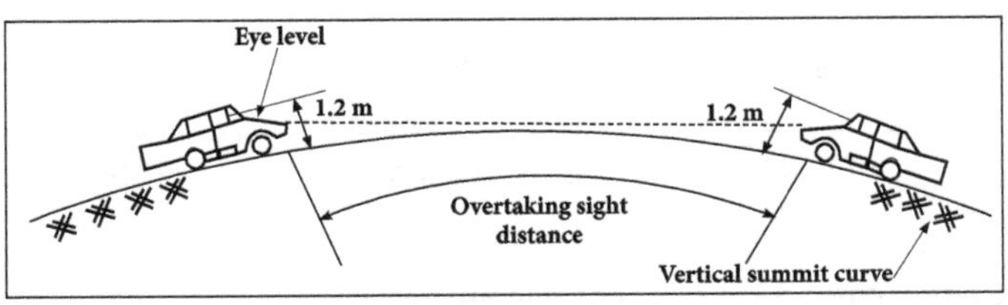

Measurement of overtaking sight distance.

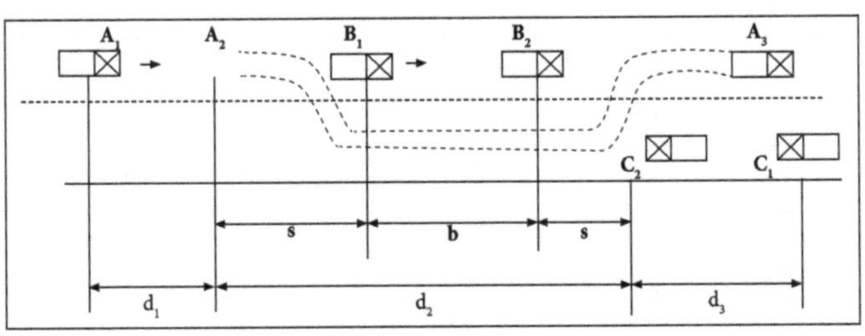

Manoeuvre of overtaking vehicle.

Minimum overtaking sight distance is governed by the following factors:

1. Speeds of overtaken and overtaking vehicles.
2. Other vehicles coming from the opposite direction.
3. Space between overtaken and overtaking vehicles.
4. Skill and reaction time of the driver.
5. Rate of acceleration of overtaking vehicle.
6. Slope of the road.

Table: Maximum overtaking acceleration at different design speeds.

Design speed	Maximum overtaking acceleration		
V kmph	v m/sec	A kmph/sec	a m/sec^2
25	6.93	5.0	1.41
30	8.34	4.8	1.30
40	11.10	4.45	1.24
50	13.86	4.00	1.11
65	18.00	3.28	0.92
80	22.20	2.56	0.72
100	27.80	1.92	0.53

The IRC has specified safe overtaking sight distances for various design speeds.

The OSD values thus obtained are rounded off and presented by IRC as given in table.

Table: Overtaking sight distance for various design speeds.

Design speed, kmph	Time components, seconds			Safe overtaking sight distance, m
For overtaking manoeuvre		For oncoming vehicle		Total
40	9	6	15	165
50	10	7	17	235
60	10.8	7.2	18	300
65	11.5	7.5	19	340
80	12.5	8.5	21	470
100	14	9.0	23	640

Overtaking Zones

It is always desirable to construct roads such that the visibility is available well ahead at every point of the road. It is not feasible in practice. In areas where visibility is not adequate a prohibitive board such as "No passing" or "Overtaking prohibited" may be displayed.

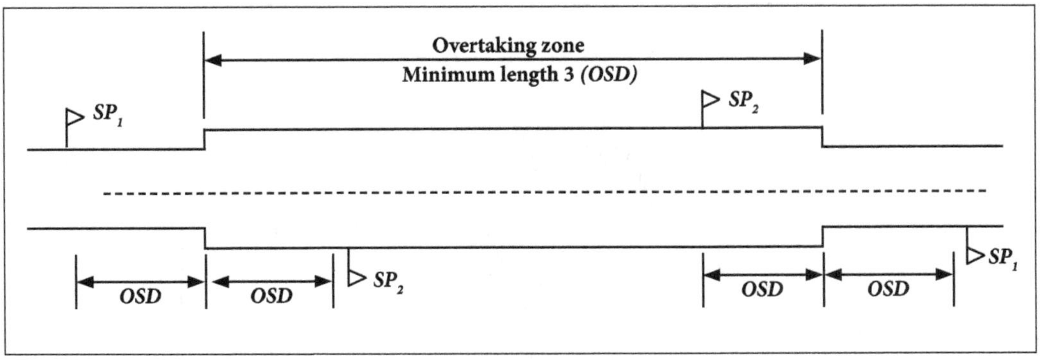

SP_1 - Sign Post "Overtaking zone ahead".

SP_2 - Sign Post "End of overtaking zone".

Vehicles moving at design speed should be given opportunity to overtake along the road at frequent intervals. Areas or zones which are meant for overtaking are called overtaking zones.

The minimum length of overtaking zone should be three times the safe over-taking distance, i.e., $3(d^1 + d^2)$ for one-way roads and $3(d^1+d^2+d^3)$ for two-way roads. It is desirable that the length of overtaking zones is kept five times the overtaking sight distance.

Adequate sign posts should be installed on either side of the overtaking zone. The sign post should be kept at a distance of (d^1+d^2) for one-way roads and $(d^1+d^2+d^3)$ for two-way roads.

Sight Distance at Intersections

Uncontrolled intersection is one where there is no automatic signal or any policeman available to control the traffic. Vehicle entering an uncontrolled intersection should have sufficient visibility so as to enable the driver to control his vehicle and to avoid collision with other vehicles moving in from other roads. It is essential that sufficient sight distance is available along the intersecting roads. Uncontrolled intersections may be grouped under two categories:

1. Uncontrolled intersections where there are intersecting roads of more or less equal importance.
2. Priority intersections consist of a main road intersecting a minor road.

1. Uncontrolled Intersection

In these intersections it is essential that there is a clear view across the corners from a sufficient distance so as to avoid collision of vehicles. Existence of some structures or some objects at the corners may obstruct the view. The unobstructed area formed by the lines of vision and sight line, called the sight triangle (ABC), is shown in figure.

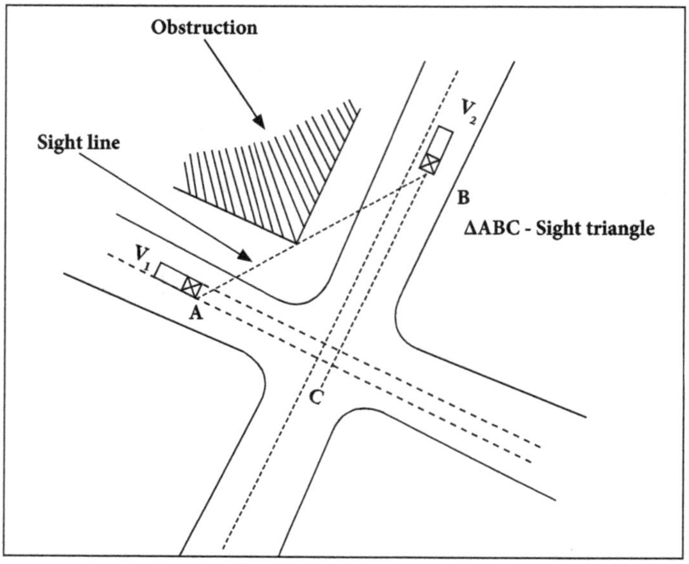

Sight distance at uncontrolled intersection.

Sight distance requirements at uncontrolled intersections may be decided on the basis of three possible conditions.

i. Vehicle to change speed.

ii. Vehicle to stop.

iii. Vehicle to cross a road.

2. Priority Intersection

Here the visibility provided should be such that the vehicles moving towards the main road are able to see the vehicles on the main or major road. The time available should be enough to judge whether the required space is available to cross the main road without any interference of the traffic on the main road. For this purpose, IRC has recommended a minimum visibility distance of 15 m along the minor road.

A driver on a minor road takes about 8 seconds to perceive the traffic conditions on the intersection, to estimate the gap in the vehicle stream, to take a decision about the crossing and to accelerate the vehicle to complete the maneuver. On this basis, the sight triangle at priority intersections should be formed with BC measuring 15 m along the minor road.

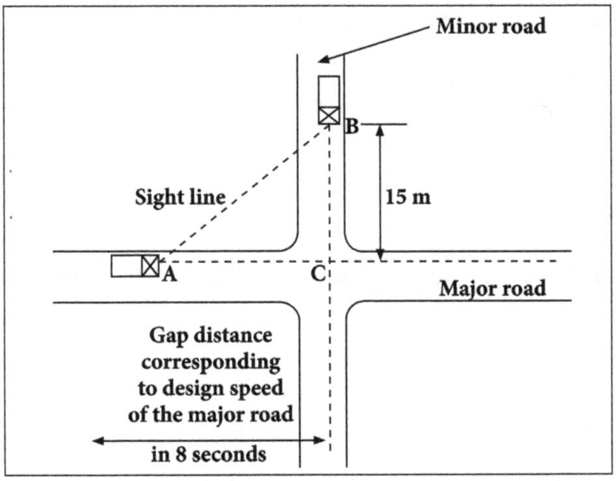

Minimum sight triangle at priority intersection.

Now the distance AC corresponds to 8 seconds reaction time and the design speed of vehicle on the major road.

Intermediate and Headlight Sight Distance

Apart from the three stopping sight distances discussed so far, the following sight distances are recommended by IRC for consideration in highway design.

1. Intermediate Sight Distance

Intermediate Sight Distance is taken as twice the Stopping Sight Distance.

2. Headlight Sight Distance

This is the visible distance for a driver during night time under the illumination of headlights.

Problem

If the acceleration of overtaking vehicle is 0.98 m/sec^2 and speed of overtaking and overtaken vehicles are 80 and 50 kmph respectively, on a two-way traffic road then with these assumptions, let us calculate:

i. Safe overtaking sight distance.

ii. Minimum length of overtaking zone and show the positions of the sign.

Solution:

Given:

Acceleration of overtaking vehicle = 0.98 m/sec^2

Speed of overtaking vehicles = 80 kmph

Speed of overtaken vehicles = 50 kmph

Formula to be used:

$$OSD = d_1 + d_2 + d_3$$

Consider the speed of the overtaking vehicle as the design speed, i.e.,

$$v = 80\,\text{kmph} = \frac{80}{3.6} = 22.22\ \text{m/sec}$$

Speed of overtaken vehicle,

$$v_b = 50\,\text{kmph} = \frac{50}{3.6} = 13.89\ \text{m/sec}$$

Acceleration, a = 0.98 m/sec^2.

$$OSD = d_1 + d_2 + d_3$$

Assume reaction time as 2 sec., i.e., t = 2 sec

Therefore, d_1 = vbt

= 13.89 × 2 = 27.78 m

s = (0.7 vb+ 6)

= 0.7 × 13.89 + 6 = 13.723 m

Time, $T = \sqrt{\dfrac{4 \times 13.723}{0.98}} = 7.6$ sec

d_2 = vbT + 2 s

= 13.89 × 7.6 + 2 × 13.723

= 127.01 m

d^3 = vT

= 22.22 × 7.6

= 161.87 m

OSD = 27.78 + 127.01 +161.87

= 316.66 m

i. i.e., OSD = 317 m.

ii. Minimum length of OSD (for two-way traffic)= 3 (OSD) = 3 × 317 = 951 m.

iii. Position of sign posts are shown in figure.

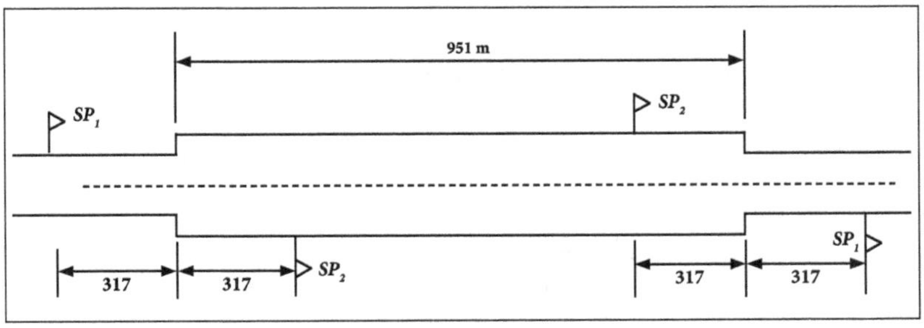

Overtaking zones

SP_1- Sign Post-Overtaking zone ahead

SP_2- Sign Post-End of Overtaking zone

2.4 Design of Horizontal Alignment

Horizontal Curves

It is provided in the horizontal plane connecting two straight alignments.

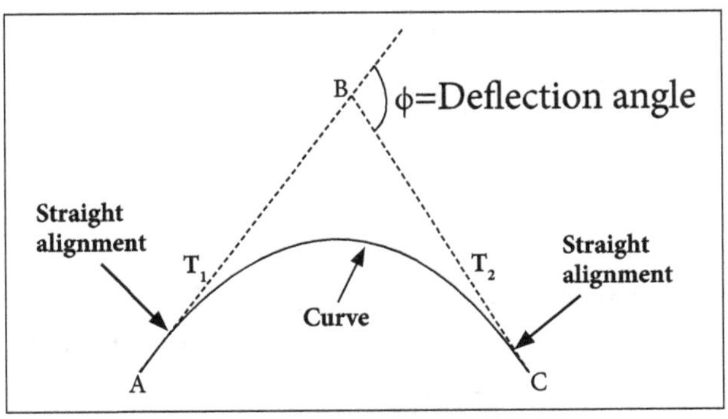

Horizontal curve

Types of Horizontal Curves

Horizontal curves may be classified as:

1. Simple curve
2. Compound curve
3. Reverse curve
4. Transition curve

Simple Curve

It consists of a single arc with a constant radius connecting two straights or tangents.

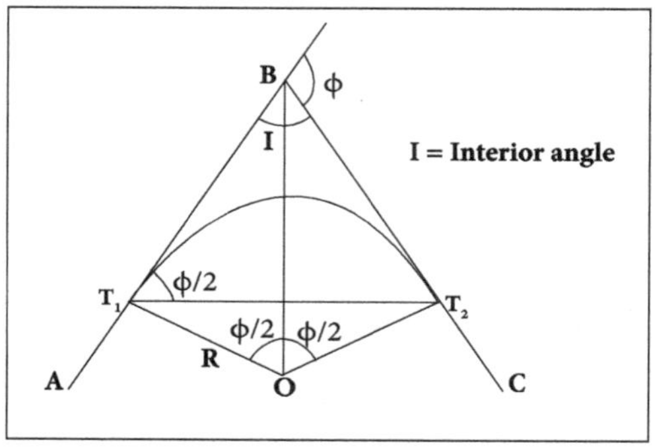

Simple circular curve.

Compound Curve

It consists of more than one radius connecting two intersecting straights.

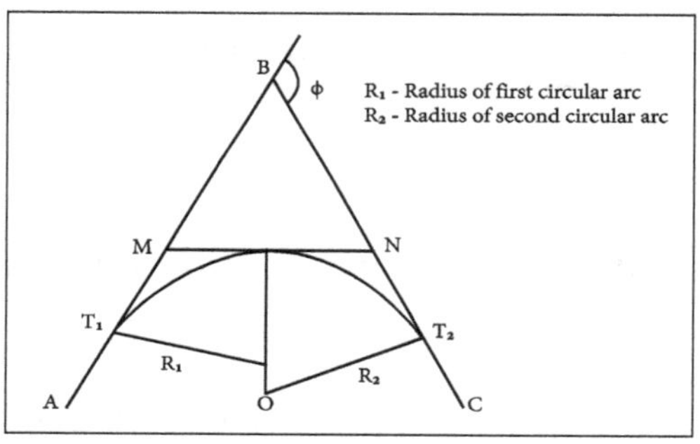

Compound curve

Reverse Curve

When two curves of different or equal radius are bending in opposite directions then it is called a reverse curve. Reverse curves have one common tangent.

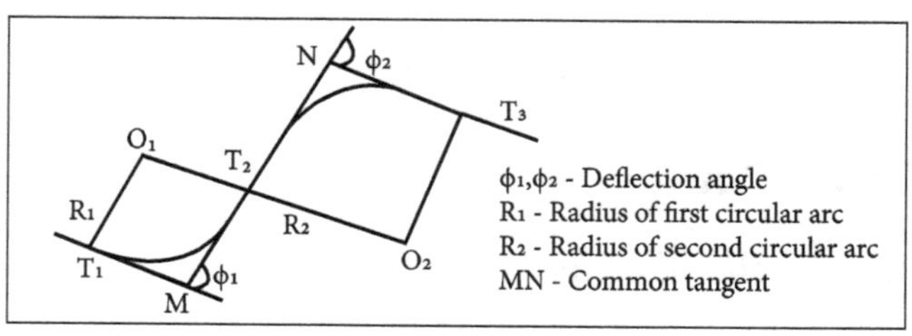

Reverse curve

Transition Curve

It is a curve having a gradual varying radius. Example, a curve with infinite radius in the beginning and varying gradually to a finite radius.

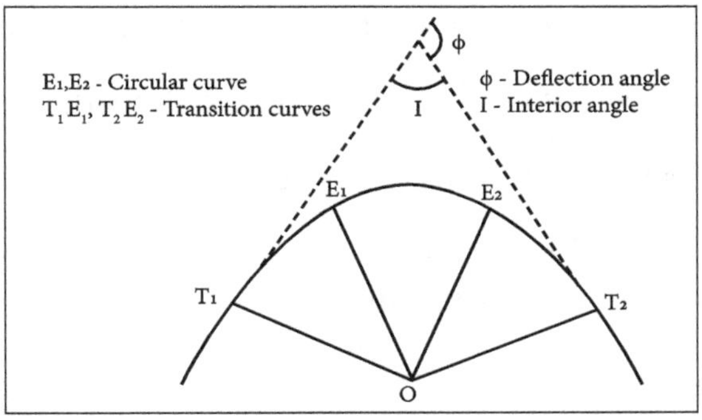

Transition curve

2.4.1 Design of Superelevation and Extra Widening

Superelevation

To counter balance the centrifugal force,the outer edge of the road is raised which is known as the superelevation or cant or banking. Transverse slope is provided throughout the length of the horizontal curve.

It is expressed as the ratio of the height of the outer edge with respect to the horizontal width. Figure shows the details of superelevation.

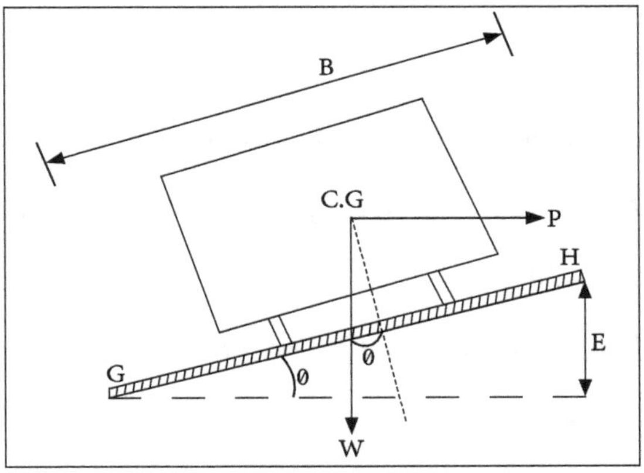

Super elevated section of pavement.

$$e = \frac{HI}{GI} = \tan\theta.$$

The value of e is very small and of the order of 0.07, hence tan θ = sin θ, then,

$$e = \tan\theta = \sin\theta = \frac{HI}{GI} = \frac{E}{B}$$

which is the height to width ratio of the pavement.

Total rise in outer edge of the pavement with respect to the inner edge HI=E = eB.

where, e is the superelevation rate and E is the total super elevated height of outer edge.

Now all the forces acting are summarised below along with the components.

i. The centrifugal force, $P = \frac{Wv^2}{gR}$, acting horizontally through the C.G.

ii. The weight W of the vehicle acting vertically downwards through the C.G.

iii. The friction developed between the wheels and the pavement, FA and FB acting parallel to pavement elevated position.

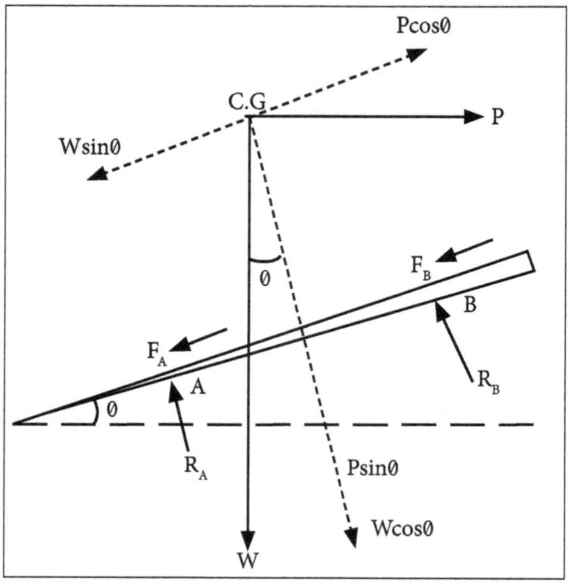

Forces acting on a moving vehicle on a superelevated pavement.

Considering the equilibrium of forces acting parallel to the pavement,

$$P \cos \theta = W \sin \theta + Fa + Fb$$

Considering the equilibrium of forces acting perpendicular to the pavement,

$$RA + RB = W \cos\theta + P \sin \theta$$

Rewriting the above equation,

$$P \cos \theta = W \sin\theta + f\,(RA + RB)$$

Substituting for RA + RB from above equation we have,

$$P \cos \theta = W \sin\theta + f\,(W \cos\theta + P \sin\theta)$$

i.e., $P\,(\cos \theta - f \sin\theta) = W \sin\theta + f\,W \cos\theta$

Dividing by W cos θ throughout,

$$\frac{P}{W}(1 - f \tan\theta) = \tan\theta + f$$

i.e., $$\frac{P}{W}=\frac{\tan\theta+f}{1-f\tan\theta}$$

The coefficient of lateral friction is generally taken as 0.15 and tan θ seldom exceeds 1/15. Hence, f tan θ is less than 0.01. Thus, the denominator of the centrifugal ratio 1-f tan θ = 0.99 which can be approximated to 1, then

$$\frac{P}{W}=\tan\theta+f=e+f$$

But $$\frac{P}{W}=\frac{v^2}{gR}$$

i.e., $$e+f=\frac{v^2}{gR}$$

The above equation can be represented taking V in kmph as,

$$e+f=\frac{(0.278V)^2}{9.8R}=\frac{V^2}{127R}$$

Neglecting the coefficient of friction, the equilibrium superelevation required to counteract the centrifugal force fully is given as,

$$e=\frac{v^2}{gR}$$

$$e=\frac{v^2}{127R}$$

Once this superelevation is provided,the pressure on the inner and outer wheels will be equal. This condition will result in a very high value of superelevation. As lateral friction cannot be ignored completely f is always taken into account.

In some field situations it may not be possible to provide superelevation (i.e., e = 0), then the later friction alone should counteract the centrifugal force. In such situations:

$$f=\frac{v}{gR}$$

or $$v=\sqrt{fgR}$$

and $$f=\frac{V^2}{127R}$$

or $$V=\sqrt{127fR}$$

1. Limits of Superelevation

i. Maximum

Amount of superelevation increases with increase in speed and with decrease in radius of the curve, for a constant value of f. Maximum e should be fixed on roads with mixed traffic.

Indian Roads Congress has fixed a maximum limit of superelevation as 7% in plain and rolling terrain and with mixed traffic. A maximum of 10% is allowed on hill roads.

ii. Minimum

From drainage point of view a certain superelevation has to be provided. If the calculated e is less than or equal to camber of the road surface thene may be limited to camber of the road on the curve.

Table: List of Radii Beyond which Superelevation is not required.

Design speed (kmph)	Radius (metres) for camber of				
	4%	3%	2.5%	2%	1.7%
20	50	60	70	90	100
25	70	90	110	140	150
30	100	130	160	200	240
35	140	180	220	270	320
40	180	240	280	350	420
50	280	370	450	550	650
65	470	620	750	950	1100
80	700	950	1100	1400	1700
100	1100	1500	1800	2200	2600

The IRC has recommended for various speed of vehicles, the radii along with camber for which no superelevation is needed.

Design of Superelevation

In the design of superelevation the following factors are to be considered:

i. For a mixed traffic, the superelevation provided for the speedy vehicle with f = 0 is safer for most of the vehicles. This may be inconvenient for slow moving vehicles.

ii. For a mixed traffic, providing superelevation based on lateral friction would be unsafe for fast moving vehicles.

iii. Considering both these aspects, as a compromise, the centrifugal force due to 75% of the design speed and limiting the maximum superelevation to 0.07 are adopted.

For hilly roads, the maximum superelevation may be taken as 0.10. Following steps are to be adopted in the design of a superelevation for a pavement:

Step 1

Superelevation for 75 % of design speed (v m/sec or V kmph) is calculated neglecting the friction (i.e., taking f = 0).

$$e = \left(\frac{(0.75v)^2}{gR} \right)$$

or $$e = \frac{(0.75v)^2}{127R} = \frac{V^2}{225R}$$

Step 2

If the computed value of e is less than 7%, then computed value is provided.

If the value is greater than 7%, then provide 7% superelevation and step 3 and 4 are followed.

Step 3

Compute the coefficient of friction developed for the maximum value of e= 0.07 at the full value of design speed from,

$$f = \left(\frac{v^2}{8R} - 0.07 \right)$$

or $$f = \left(\frac{V^2}{127R} - 0.07 \right)$$

If the computed f value is less than 0.15, then the superelevation provided (7%), is safe for the design speed. If not, the restricted speed limiting f = 0.15, is computed as explained in Step 4.

Step 4

The allowable speed for the given curve for the maximum superelevation of 7% and friction of 0.15, that is

$$e + f = 0.07 + 0.15 \div 0.22$$

i.e., $$e + f = \frac{v_a^2}{8R}$$

or $$e + f = \frac{V_a^2}{127R}$$

If the calculated allowable speed is higher than the design speed then the design is adequate and the superelevation, e = 0.07 is provided.

If the calculated allowable speed is less than the design speed, the speed is limited to V_a kmph as calculated above. Speed limit signs are installed at appropriate places.

In the case of important highways, speed restrictions cannot be made. Instead the curve should be re-aligned, if adequate land is available, with a large radius so that the design speed can be maintained.

Widening of Pavement on Horizontal Curve

In horizontal curves, specially when the radii of the curves are low, it is the practice to widen the pavement width a little extra for safe driving.

Reasons for Widening

i. Majority of vehicles have the facility to turn only the front wheels and have a rigid wheel base. Jeep like vehicles have a fourwheel drive facility. When vehicles with twowheel drive facility take a turn to negotiate a horizontal curve, the rear wheels do not follow the same path on that of the front wheels. This is referred to as offtracking.

ii. At high speeds when adequate superelevation and lateral friction are not available, there will be some transverse skidding and the rear wheel may take paths on the outside of those traced by the front wheels.

iii. The drivers may have a tendency, to follow the curved path with larger radius and to have a greater visibility at curve.

iv. While negotiating an oncoming vehicle or overtaking a horizontal curve there is a psychological tendency for the drivers to adopt a greater clearance between the vehicles.

To take care of these aspects, it has been the practice to widen the pavement on the horizontal curves when the radius of the curve is less than 300 m.

The extra widening needed comprises of two parts, viz.,

i. Mechanical widening

ii. Psychological widening

1. Mechanical Widening

Mechanical widening is provided to account for offtracking due to the rigidity of the wheel base. The movement of the vehicle is shown in figure.

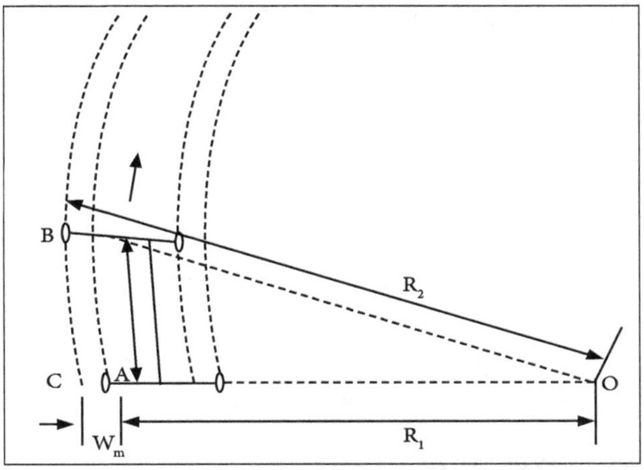

Mechanical widening on horizontal curve.

Where,

R_1 - Radius of the path traversed by the outer rear wheel, metres

R_2 - Radius of the path traversed by the outer front wheel, metres

W_m - Mechanical widening

l -Length of the wheel base, metre

2. Psychological Widening

The psychological widening is mainly provided for greater manoeuvrability of the vehicle. This is particularly important in pavements with more than one lane.

An empirical formula has been recommended by IRC for psychological widening, W_{ps}, as

$$W_{ps} = \frac{V}{9.5\sqrt{R}}$$

3. Total Widening

Total extra width W_e provided for a load with n lanes is given as,

$$W_e = \frac{nl^2}{2R} + \frac{V}{9.5\sqrt{R}}$$

where,

R = Mean radius of the curve.

l = Length of wheel base of longest vehicle (m).

The extra width recommended by the Indian Roads Congress for single and two lane pavements are presented in table.

Table: Extra width of pavement at horizontal curves.

Radius of curve (m) Extra width (m)	up to 20	21 to 40	44 to 60	61 to 100	101 to 300	Above 300
Single-Lane	0.9	0.6	0.6	Nil	Nil	Nil
Two-Lane	1.5	1.5	1.2	0.9	0.6	Nil

Note: For multi-lane roads, the pavement widening is calculated by adding half the extra width of two-lane roads to each lane of the multi-lane road.

4. Introduction of Extra Widening

Widening is introduced gradually starting from the tangent point (T.P). If the curve is provided with a transition curve, then it is started from the beginning of the transition curve and provided with uniform rate till the full value of We is reached. At this point the full superelevation is provided. The widening is provided equally on either side of the curve at every cross section up to a value of $W_e/2$.

On horizontal curves without transition curve, two-thirds of the widening is provided at the end of the straight section and one-thirds is provided on the circular curve beyond the tangent point Here, widening is provided only on the inside of the curve.

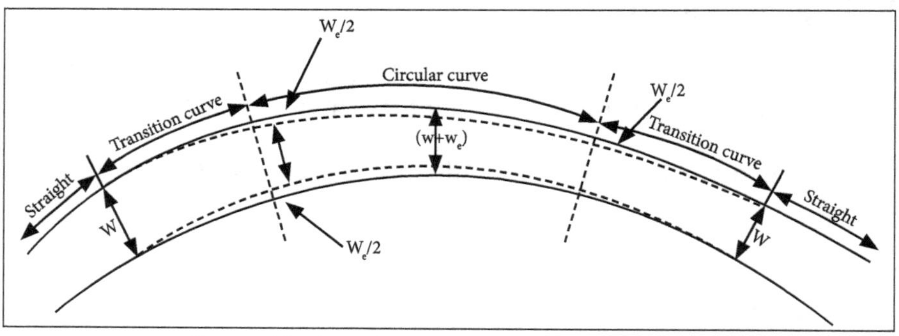

Extra widening of pavement on horizontal curve.

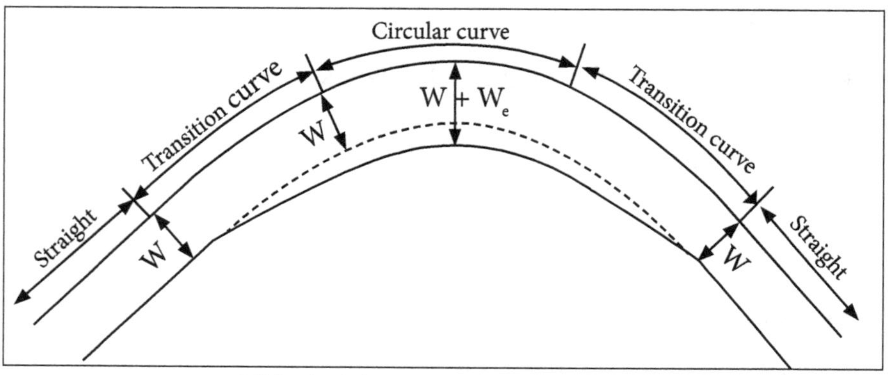

Widening of pavement on sharp curve.

Problem

The design speed on a road with a horizontal curve of radius 400 m is 90 km/hr. The coefficient of friction available on the road may be taken as 0.15. Let us calculate:

i. Superelevation when full lateral friction has developed.

ii. Coefficient of friction when no superelevation is provided.

iii. Superelevation for equal pressure at inner and outer wheels.

Solution:

Given:

Radius = 400 m

Design speed = 90 km/hr

Coefficient of friction available on the road = 0.15

Formula to be used:

$$e+f=\frac{V^2}{127R}$$

i. We know that,

$$e+f=\frac{V^2}{127R}$$

Substituting the known values,

$$e+0.15=\frac{(90)^2}{127\times 400}=0.1595$$

$$e = 0.1595-0.15 = 0.0095$$

i.e., Superelevation rate = 0.95 %

ii. No superelevation is provided, i.e., e = 0

Then, coefficient of friction, $f=\frac{V^2}{127R}=\frac{90^2}{127\times 400}=0.1595$

iii. When pressure on the inner and outer wheels is equilibrium condition, superelevation for which f = 0

$$\therefore \quad e+0.15=\frac{(90)^2}{127\times 400}=0.1595$$

Superelevation is 15.95% which is more than the maximum permissible percentage of 7% and hence cannot be provided.

2.4.2 Design of Transition Curves

Transition Curve is also called as an easement curve. It is an arc introduced between a straight and a circular curve or between two arcs of a compound curve. The radius of a transition curve varies from infinity to a fixed value.

Vehicles passing straight into a circular curve experience jerk due to the effect of centrifugal force. Centrifugal force always acts in a direction perpendicular to the axis of rotation.

To counter act the above effect, the outer edge of the road or the rail is raised which is called the superelevation or cant. Superelevation is provided gradually along the transition curve.

Advantages

i. Introduces superelevation gradually from zero at the tangent point to the value on the circular curve.

ii. Maintains a constant proportional between superelevation and rate of change of curvature.

iii. Eliminates discomfort, overturning and side slipping of vehicles.

iv. Eliminates discomfort to passengers.

Requirements

A transition curve should satisfy the following requirements:

i. The transition should be tangential to the straight curve.

ii. The curvature of the transition curve should be zero at the origin of the straight curve.

iii. The exact amount of superelevation should be attained at the junction with the circular curve.

iv. The curvature of the transition curve should increase at the same rate as that of the superelevation.

Details of a Composite Curve

Transition curves are introduced at both ends of a circular curve, the resulting curve is known as a combined or composite curve.

AB - The rear or back tangent.

BC - The forward tangent.

T_1 - Starting point of the curve or first tangent.

T_2 - Ending point of the curve or second tangent.

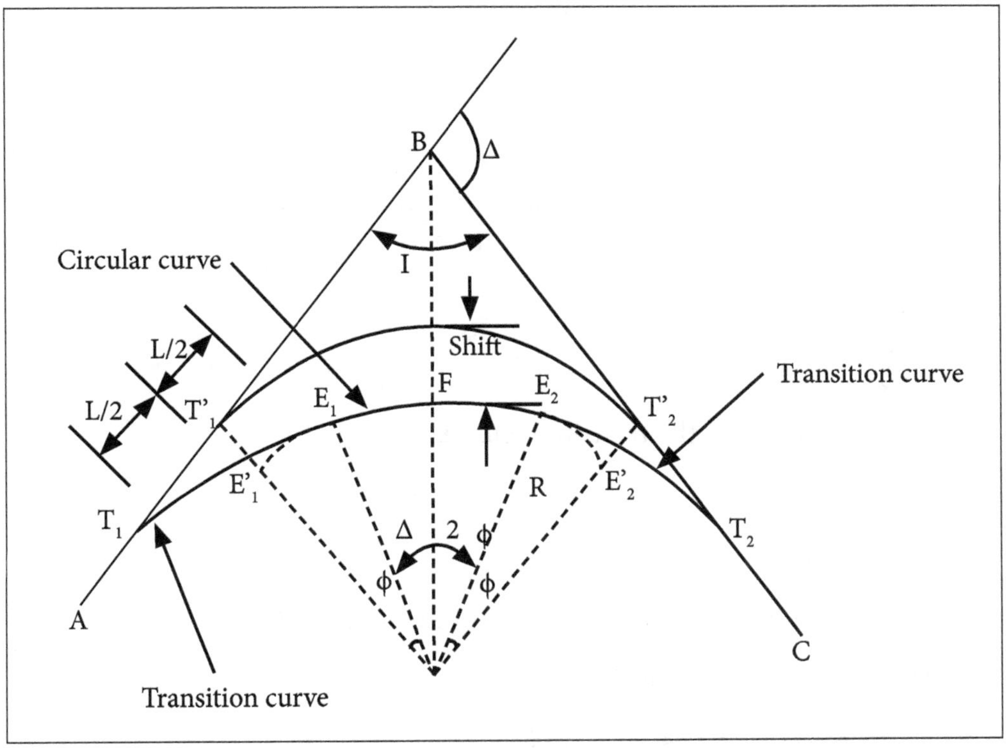

Composite curve

D - Deflection angle

I - Angle of intersection

φ - Spiral angle

T_1E or T_2D - Length of the transition curve

EFD - Length of circular curve

$T_1'E'_1$ or $T_2'E'_1$ - Shift of the curve

BT_1 or BT_2 - Total tangent length

$T_1E\ F\ DT_2$ - Length of the combined curve

Angle (Δ - 2φ) - Central angle

Types of Transition Curves

i. Euler's Spiral

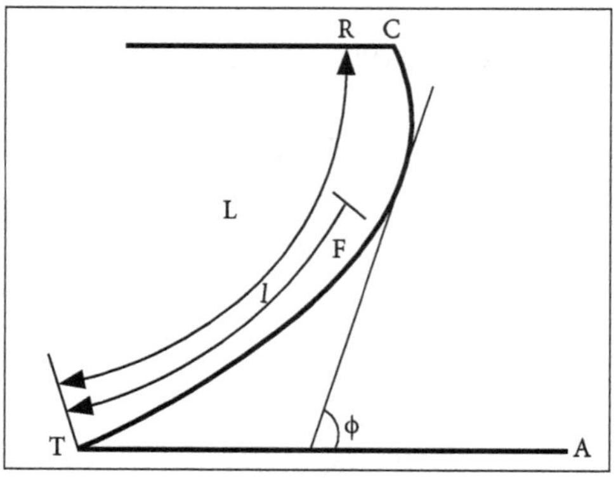

Euler's spiral

Equation for the above curve,

$$\phi = \frac{l^2}{2RL}$$

This is the ideal equation of the transition curve.

Here,

F - A point on the curve

φ - Angle made by tangent at F with initial tangent TA

l - Distance between F and T

L - Total length of transition curve

R - Radius of circular curve

ii. Cubical Spiral

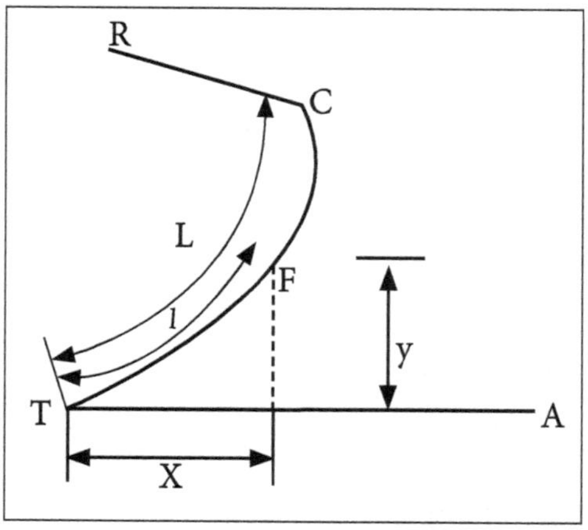

Cubical spiral

The equation of this curve is

$$y = \frac{l^2}{6RL}$$

Where,

y - Perpendicular offset to any point from TA

l - Distance of the point from T

L - Total length of transition curve

R - Radius of circular curve

iii. Cubic Parabola

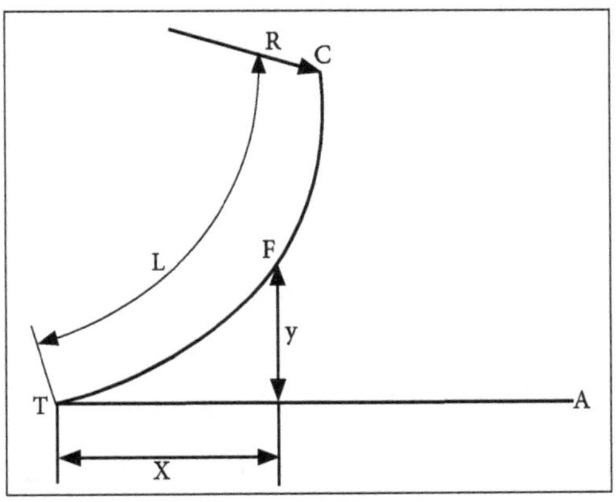

Cubic parabola

Figure shows a cubic parabola.

$$y = \frac{x^3}{6RL}$$

Where,

x - Horizontal distance between F and T.

y - Perpendicular distance between F and TA.

Cubic parabolas are generally recommended for railways.

Details of Transition Curve

i. Length

Length of a transition curve is calculated based on one of the following considerations:

1. By adopting a particular rate of superelevation.
2. By considering arbitrary rate of superelevation.

3. By considering rate of change of radial acceleration.

By adopting a particular rate of superelevation, the definite rate of superelevation varies from 1 in 300 to 1 in 1000.

Length of transition curve,

$$L = \frac{ne}{100} m$$

where,

e - Amount of superelevation in centimeters.

n - Rate of superelevation.

By considering arbitrary rate of superelevation,

Length of transition curve,

$$L = \frac{e \times V}{x}$$

Where,

e - Amount of superelevation

V - Speed in meters per second

x - Time rate in centimeters per second (varies from 2.5 to 5 cm/sec)

By considering rate of change of radial acceleration,

Radius acceleration on circular curve = $\frac{V^2}{R}$

Time taken by the vehicle to cover transition curve = $\frac{L}{V}$ s

If the change of radial acceleration is K m/s^3 then,

Time taken to attain maximum radial acceleration = $\frac{V^2}{KR}$ s

From equation,

$$\frac{L}{V} = \frac{V^2}{KR}$$

i.e., $$L = \frac{V^3}{KR}$$

ii. Spiral Angle

From the intrinsic equation of the ideal transition curve, we have

$$\phi_1 = \frac{l^2}{2RL} \text{radians}$$

The angle ϕ is known as the spiral angle.

iii. Deflection Angle

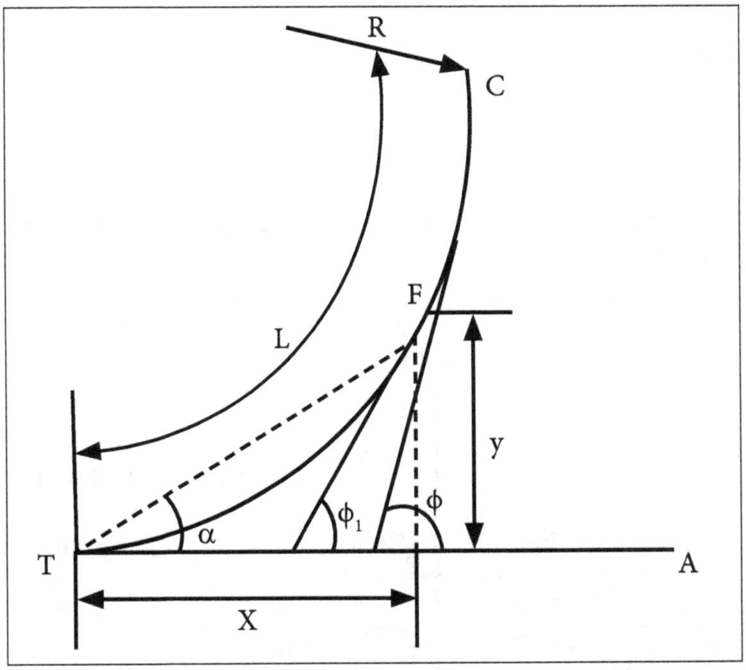

Deflection angle

Here,

F - Any point on transition curve

C - Point of junction of circular and transition curve

ϕ - Spiral angle

ϕ - Angle between tangent at F and initial tangent

α - Deflection angle at F

x and y - Rectangular coordinates

R - Radius of circular curve

i - Length of transition curve between T and F

Now, $\tan\alpha = \frac{y}{x}$

From equation for cubical spiral, we know that,

$$y = \frac{I^3}{6RL}(\text{where } x \approx 1)$$

Considering α as small, then tan α ≈ α

$$\therefore \alpha = \frac{1}{x} \times \frac{x^3}{6RL} = \frac{x^2}{6RL} = \frac{l^1}{6RL} \text{radians}$$

$$\alpha = \frac{573l^2}{RL} \text{mins}$$

iv. Shift

The distance by which the main curve is moved inwards to accommodate the transition centre is called a shift.

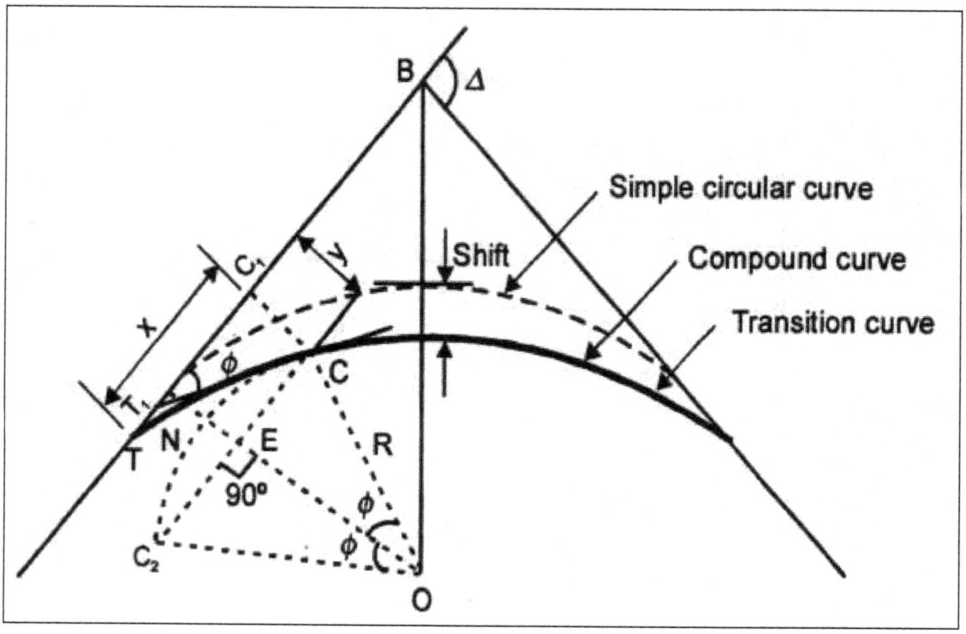

Compound curve

Here,

TB - Initial tangent

T - Tangent point of combined curve

T_1 - Tangent point of simple circular curve

S - Shift

C - Junction point of transition curve with circular curve

DC - Common tangent

ϕ - Spiral angle

N - Intersection point of shift and transition curve

R - Radius of circular curve

L - Length of transition curve

From the intrinsic equation of the ideal transition curve, we have

Spiral angle,

$$\phi = \frac{L}{2R}$$

$$CE = R\phi = R \times \frac{L}{2R} = \frac{L}{2}$$

But CN is very nearly equal to CE.

$$CN = \frac{L}{2}$$

Hence, $TN = \frac{L}{2}$

So the shift T_1E bisects the transition curve at N.

Now, CC_2 is perpendicular to OT_1 and CC_2 is perpendicular to TB.

$$S = C_2T_1 - C_2E$$

(where C_2E = versed sine of curve with central angle 2φ)

$$= CC_1 - R(1-\cos\varphi)$$

$$= y - R(1-\cos\varphi)$$

$\therefore$ $$S = y - R \times 2\sin^2\frac{\phi}{2}$$

We know,

$$y = \frac{x^3}{6RL} = \frac{L^3}{6RL} = \frac{L^2}{6R} \quad (x \approx L)$$

$$S = \frac{L^2}{6R} - R2\left(\frac{\phi}{2}\right) = \frac{L^2}{6R} - 2R \times \frac{\phi^2}{4}$$

$$= \frac{L^2}{6R} - 2R \times \frac{1}{4}\left(\frac{L}{2R}\right)^2$$

$$= \frac{L^2}{24R}$$

Considering φ as small, sin φ ≈ φ

i.e., Shift $S = \frac{L^2}{24R}$

Again $T_1N = \frac{TN^3}{6RL} = \frac{(L/2)^2}{6RL} = \frac{L^2}{48R}\left(TT_1 = TN = \frac{L}{2}\right)$

$$\therefore \quad \frac{\text{Shift}}{T_1N} = \frac{L^2N}{24R} \times \frac{48R}{L^2} = 2$$

Hence, T_1N=1/2 shift = 1/2 T_1E

v. Elements of Combined Curve

a. The length of the transition curve may be calculated from the formula, if not given,

$$L = \frac{V^3}{KR}$$

$$S = \frac{L^2}{24R}$$

b. Shift,

where R = Radius of circular curve.

c. Tangent length of the combined curve (cubic parabola as transit curve)

$$BT'_1 = BT_1 + T_1T'_1 = OT_1 \tan\frac{\Delta}{2} + \frac{L}{2}$$

$$= (R+S)\tan\frac{\Delta}{2} + \frac{L}{2}$$

where Δ = Deflection angle

d. Spiral angle

$$\phi = \frac{L}{2R}\text{radians} = \frac{L}{2R} \times \frac{180^\circ}{\pi}\text{deg.}$$

e. Central angle for circular curve = Δ - 2 φ

f. Length of circular curve = $\frac{\pi R(\Delta - 2\phi)}{180^\circ}$

g. Length of combined curve = $T_1E_1 + E_1E_2 + E_2T_2$

$$= L + \frac{\pi R(\Delta - 2\phi)}{180^\circ} + L = \frac{\pi R(\Delta - 2\phi)}{180^\circ} 2L$$

h. Chainage of point of start of the curve is given by T_1 = chainage of B - BT_1

i. Chainage of E_1 = Chainage of T_1 + (length of transition curve)

j. Chainage of E_2 = Chainage of E_1 + (length of circular curve)

k. Chainage of point of tangency,

T_2 = Chainage of E_2 + (length of transition curve)

l. Tangent length of combined curve

$$=(R+S)\tan\frac{\Delta}{2}+\frac{L}{2}\left(1-\frac{S}{5R}\right)$$

Problem

Design speed for plain terrain, V = 100 kmph

Number of lanes, n = 2

Pavement width = 7.0 m

Wheel base of truck, l = 6.0 m

Maximum value of superelevation, e = 0.07

Skid resistance, f = 0.15

With this data, let us compute the additional width required for a horizontal curve of a state highway in a plain area with a ruling minimum radius.

Solution:

Given:

Design speed for plain terrain, V = 100 kmph

Number of lanes, n = 2

Pavement width = 7.0 m

Wheel base of truck, l = 6.0 m

Maximum value of superelevation, e = 0.07

Skid resistance, f = 0.15

Formula to be used:

Therefore,
$$R_{ruling} = \frac{V^2}{127(e+f)}$$
$$= \frac{100^2}{127(0.07+0.15)}$$
$$= 357{\cdot}9$$
$$= 360\,\text{m}$$

Additional width
$$W_e = \frac{nl^2}{2R} + \frac{V}{9.5\sqrt{R}}$$
$$= \frac{2\times 6^2}{2\times 360} + \frac{100}{9.5\sqrt{R}}$$
$$= 0.10 + 0.56$$
$$= 0.66\ \text{m}$$

Total pavement width
$$= 7.0 + 0.66$$
$$= 7.66\ \text{m}$$

2.5 Design of Vertical Alignment

The vertical alignment of a road consists of gradients and vertical curves. The vertical alignment is usually drawn as a profile, which is a graph with elevation as vertical axis and the horizontal distance along the centre line of the road as the the horizontal axis.

Circular curve is used to connect horizontal straight stretches of road, vertical curves connect two gradients. When these two curves meet, they form either convex or concave. The former is called a summit curve, while the latter is called a valley curve.

2.5.1 Gradients

It is the rate of rise or fall of a road level along its length. It is expressed as a percentage rise or fall or a rate of rise or fall with respect to horizontal distance.

A gradient of 1 in 20 or 5% represents that there is an ascending or descending of road profile by one meter for every twenty metres. As the gradient on the road is not very steep the length along the road is taken as a horizontal length.

Grades are expressed as the horizontal distance along the length of the road. Longitudinal grades are almost permanent features of the highway. This is the case in urban and built up areas where no changes can be incorporated easily.

The design of a longitudinal grade depends on following factors:

i. Anticipated traffic

ii. Site conditions:

  a. Drainage

  b. Appearance

  c. Approach to adjacent property

  d. Safety aspects

iii. Railway intersections

iv. Bridges and culverts

v. Design speed

vi. Sighting line and sight distance

Types of Gradient

Gradients are classified as:

i. Ruling gradient

ii. Limiting gradient

iii. Exceptional gradient

iv. Minimum gradient

i. Ruling Gradient

It is the desirable upper limit of gradient adopted in the normal course of design. Adoption of ruling gradient should balance the cutting and filling of earth work which will give an economical design. Different factors which are to be considered in the choice of ruling gradient is:

a. Type of terrain

b. Length of the grade

c. Speed

d. Pulling capacity of vehicles

e. The presence of horizontal stretches of road

Indian Roads Congress has recommended gradients for roads in different terrains which are specified in the table below.

Table: Gradients for roads in different terrains (CRC).

Type of terrain	Rolling gradient	Limiting gradient	Exceptional gradient
Plain or rolling	3.3% (1 in 30)	5.0 % (1 in 20)	6.7% (1 in 15)
Mountanious terrain and steep terrain having elevation more than 300 m above mean sea level	5.0 % (1 in 20)	6.0 % (1 in 16.7)	7.0% (1 in 14.3)
Steep terrain up to 3000 m height above mean sea level	6.0% (1 in 16.7)	7.0 % (1 in 14.3)	8.0% (1 in 12.5)

ii. Limiting Gradient

It is provided when topography is not permitting to provide the ruling gradient, due to economy. Length of continuous grade of stretch of road steeper than ruling value should be limited. It happens in rolling and hilly terrains to go in for limiting gradient but it has to be separated by straight stretches or road with easier grade frequently.

iii. Exceptional Gradient

In some ground conditions it will be inevitable to provide a gradient steeper than limiting gradients such gradients are referred to as exceptional gradient. Such situations may arise in approaches to causeways, near hair-pin bends, etc. In a hilly terrain short stretches of grades, steeper than the maximum gradient may have to be resorted to reach an obligatory point on a fixed route.

The exceptional gradients have several adverse effects, viz., more fuel consumption, more friction losses, reduces engine efficiency, increased wear of road surface, etc. Hence, it should be adopted only in very essential field conditions.

iv. Minimum Gradient

A road with less gradient or level may not be in a position to drain easily. The surface water may be drained to the side drain due to the camber. A longitudinal slope is needed to drain the water on the surface and from the side drains.

It is essential to have a certain minimum gradient on roads from drainage point of view provided topography permits. The minimum gradient depends on the factors like rainfall, runoff, type of soil, topography and other site conditions.

Grade Compensation

In addition to gradient, if there is horizontal curve then there will be added resistance to traction. In such cases, the total resistance due to grade and curve should not exceed the resistance due to the maximum value of the gradient specified.

For design purposes, the maximum gradient may be considered as that of ruling gradient. In some special cases, the limited gradient may be used.

The gradient should be reduced in situations where sharp horizontal curves are present so as to compensate for the loss of tractive effort due to the curve. This reduction in gradient at the horizontal curve is called the grade compensation. It is provided to offset the extra tractive effort involved at the curve.

Grade compensation (%) = $\frac{30+R}{R}$ subject to maximum value of 7.5 / R,

where,

R is the radius of the circular curve in metres.

As per IRC, there is no need for grade compensation for gradients flatter than 4.0%.

2.5.2 Vertical Curves

In a highway a change in the rate of grade or direction may cause the vehicle a sudden impact. For a smooth and safe running of vehicles, the change in gradient or direction is smoothened by a curve called a vertical curve. Vertical curve is used to smooth out intersecting grade or change in direction of roads.

A parabolic curve provides a smooth and most comfortable ride as the rate of change of grade in a parabola with a vertical axis is uniform throughout. The vertical curves adopted in highway may be classified as:

i. Summit curve or crest curve with convexity.

ii. Valley or sag curve with concavity upwards.

Summit Curve

A curve with convexity upwards is called a summit curve. This occurs when,

a. An ascending gradient meets another ascending gradient.

b. An ascending gradient intersects a descending gradient.

c. A descending gradient meets another descending gradient.

d. An ascending gradient meets a horizontal.

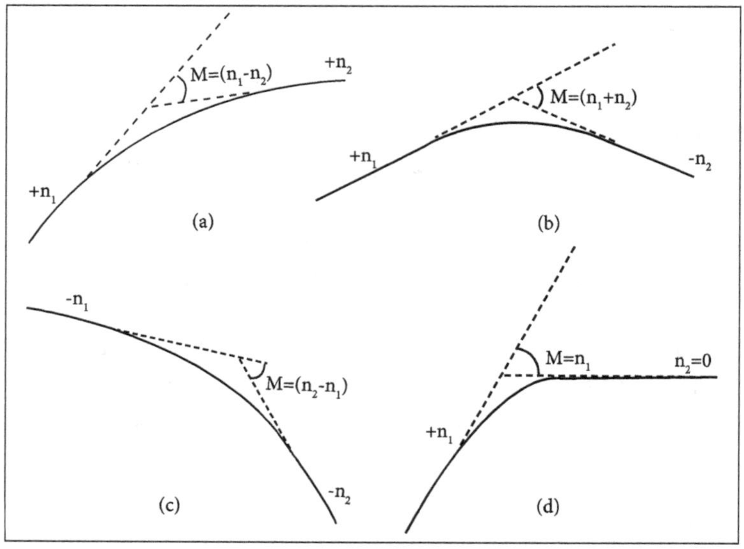

Types of summit curves.

Suitability of Curve for Summit Curve

Circular curve is ideal for a summit curve as the sight distance available throughout the curve is constant. A transition curve is undesirable as the radius of curvature changes and hence the sight distance will vary from point to point. The deviation angles in vertical curves are very small.

As a result, tangent points of a circle and a parabola are congruent. A parabola is very easy for arithmetical manipulation for computing ordinates. Use of simple parabola as summit curve is found to give good riding comfort.

Length of Summit Curve

The equation for parabolic summit curve is given by,

$$y = ax^2$$

where,

$$\alpha = \frac{N}{2L}$$

N = Deviation angle.

L = Length of the curve.

As the summit curves are long and flat, the length of the curve is taken as its horizontal projection, e.g., L = AJ as shown in figure.

Stopping Sight Distance (SSD) and Overtaking Sight Distance (OSD) should be separately considered for designing the length of the parabolic summit curve.

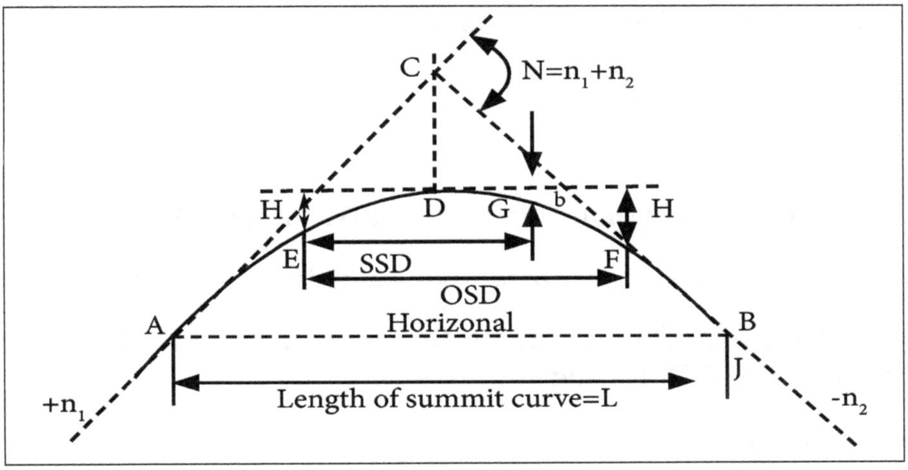

Details of a summit curve.

Length of Summit Curve for Stopping Sight Distance

Two cases have to be considered.

i. Length of summit curve is greater than the Stopping Sight Distance. (L > SSD)

ii. Length of summit curve is less than the Stopping Sight Distance (L < SSD).

i. L > SSD

The general equation for length L of the parabolic curve is given as,

$$L = \frac{NS^2}{\left(\sqrt{2H} + \sqrt{2h}\right)^2}$$

where,

L = Length of summit curve, metres

S = Stopping sight distance.

N = Deviation angle (equal to algebric difference in grades)

H = Height of eye level of driver above the road surface, metres

h = Height of subject above the road surface, metres

The value of H is 1.2 m in India and h is 0.15 m as per IRC. Substituting these values in the above we get,

$$L = \frac{NS^2}{4.4}$$

ii. L < SSD

For the condition that L is less than SSD is given as,

$$L = 2S - \frac{\left(\sqrt{2H} + \sqrt{2h}\right)^2}{N}$$

In order to find the length of summit curve, the SSD is calculated or taken from table.

Length of summit curve from above two equations by trial is used to get the required length.

Length of Summit Curve for Safe Overtaking Sight Distance or Intermediate Sight Distance

Here again two cases are to be considered:

i. (L > S) and

ii. (L < S)

Where,

L - Length of the summit curve

S - Overtaking sight distance or intermediate sight distance.

i. L > S

The general equation is applicable in this case also.

In this case h = H = 1.2 m. Substituting h = H, we get,

$$L = \frac{NS^2}{8H}$$

i.e., $$L = \frac{NS^2}{8 \times 1.2} = \frac{NS^2}{9.6}$$

ii. L < S

Here again substituting h = H = 1.2 m in the general equation,

$$L = 2S - \frac{9.6}{N}$$

In both the cases the minimum radius of the curve R is calculated from the relation R = L/N.

When the deviation angle is small, the length of summit curve works out less than the sight distance.

When the deviation angle is very small, a negative value for length of summit curve may be obtained which indicates that there is no sight restriction at the summit curve.

Minimum length of vertical curve for different speeds and for the maximum grade change which do not need vertical curve is shown in table.

Table: Requirement of grade change.

Design speed (kmph)	Maximum grade change (percentage)not requiring a vertical curve	Maximum length of vertical curve (metres)
35	1.5	15
40	1.2	20
50	1.0	30
65	0.8	40
80	0.6	50
100	0.5	60

The highest point on the summit curve is a distance Ln^N from the tangent point on the first grade n_1.

Valley Curves

A vertical curve, concave upwards, is called a valley or sag curve. This occurs when,

1. A descending gradient meets another descending gradient.
2. A descending gradient intersects an ascending gradient.
3. An ascending gradient meets another ascending gradient.
4. A descending gradient meets a horizontal.

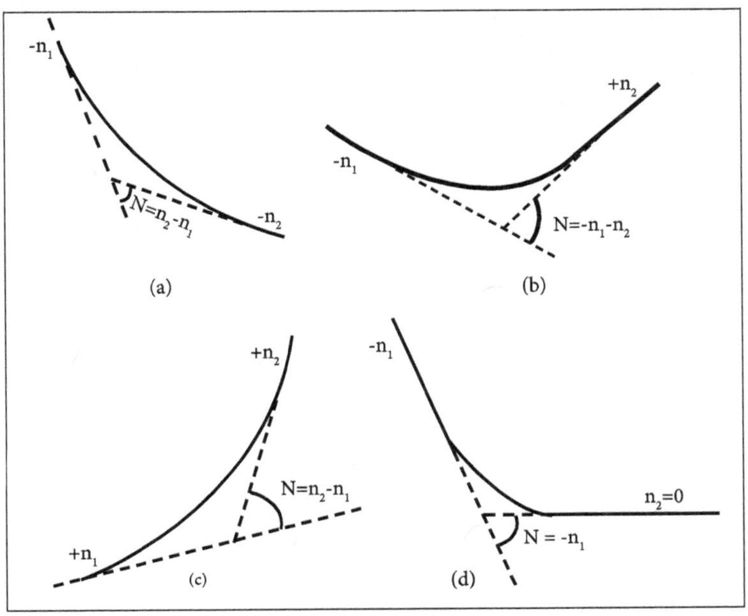

Types of valley curves.

Suitability of Curve for Valley Curve

In a valley curve, a large component of the centrifugal force will act vertically downwards. It exerts extra pressure on the tyres and springs of the vehicle. Sudden entry of the vehicle from a straight path (R is infinity) to a curved path of fixed curvature causes a shock and discomfort. Centrifugal force varies with the square of the speed and therefore may cause trouble if the roads are not planned.

It is necessary that the centrifugal force should be gradually applied. To satisfy this, the property of the curve should be such that its radius of curvature decreases gradually as its length increases. Hence transition curve is a most suitable of the three transition curves, viz., spiral, cubic parabola and the lamniscate, Cubic parabola satisfies the requirements for the valley curve.

Length of Valley Curve

The length of a valley curve is designed on the basis of two criteria:

i. The allowable rate of change of centrifugal acceleration is 0.6 m/sec^2.

ii. The headlight sight distance.

High value of length based on the above two criteria is adopted. Second criterion also gives the higher value. Figure shows the transition curve, ABC with AB and BC are two equal transition curves with, $L_s = L/2$, having the minimum radius R at the common point B.

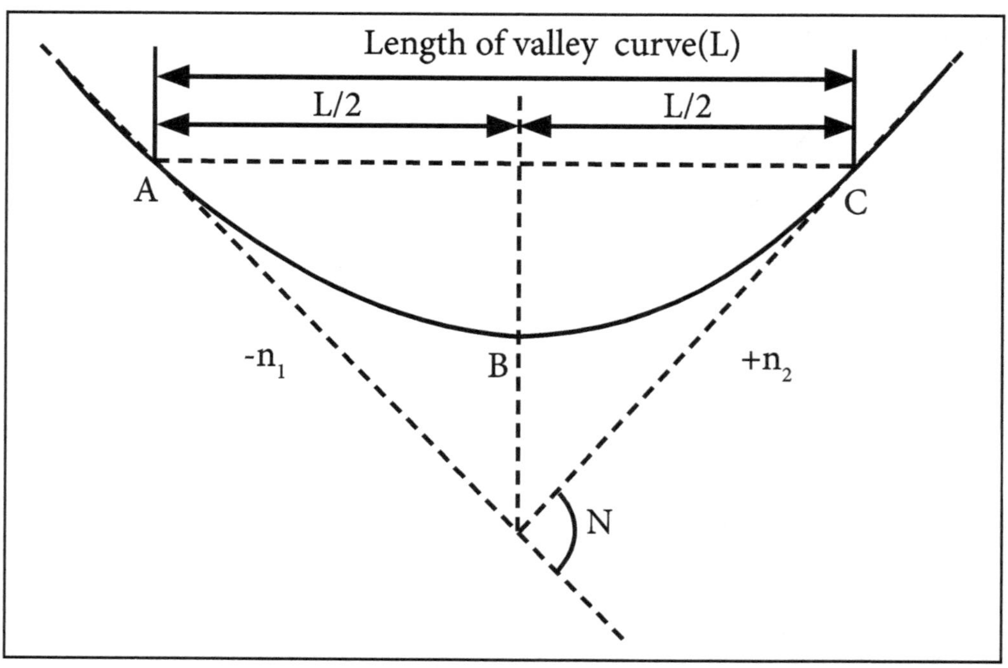

Details of valley or sag curve.

i. For Comfort Condition

The length of transition curve L for comfort condition is given as,

$$L_S = \frac{v^3}{CR}$$

Value of R at length $L_S = L_S/N = \frac{L}{2N}$

Substituting in the above equation,

$$L_S = \frac{v^3}{CL_s} N \text{ or } L_S^2 = \frac{Nv^3}{C}$$

$$\text{i.e., } L_S = \sqrt{\left[\frac{Nv^3}{C}\right]}$$

$$\text{i.e., } L = 2L_s 2\sqrt{\left[\frac{Nv^3}{C}\right]}$$

where v = Speed in m/sec.

C = Allowable rate of change of centrifugal acceleration

Taking C = 0.6 m/sec² and taking speed in kmph

$$L_s^2 = \frac{NV^3}{0.6 \times 3.6^3}$$

$$\text{i.e., } L_S = 0.19\sqrt{(NV^3)}$$

$$\text{i.e., } L = 2L_s 0.38\sqrt{VN^3}$$

Then,

the minimum radius (R in metres) of the valley curve for cubic parabola is given as,

$$R = \frac{L_S}{N} = \frac{L}{2N}$$

For Headlight Sight Distance Condition

The length of valley curve has to be determined for the two conditions:

a. When the total length of valley curve L is greater than the stopping sight distance.

b. When L is less than SSD, as given below.

a. L > SSD

The length of valley curve L is assumed to be greater than the headlight sight distance which should be equal to SSD.

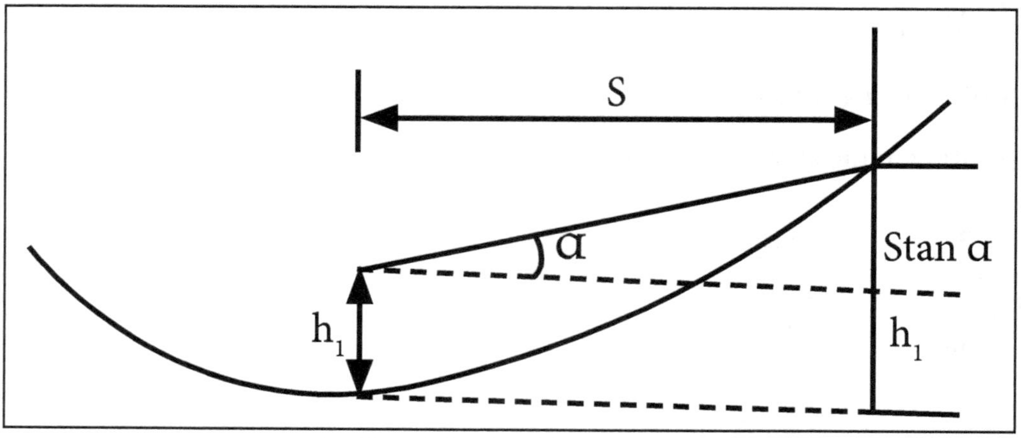

Details of headlight sight distance when L > S.

The sight distance will be minimum when the vehicle is at the lowest point on the sag curve. Assuming the valley curve as a parabola, then y =ax² and

$$h_1 + S\tan\alpha = aS^2 = \frac{NS^2}{2L} \text{(Since a = N/2L)}$$

Then, $L = \dfrac{NS^2}{2h_1 + 2S\tan\alpha}$

where, h_1 = Height of the headlight

α = Beam of headlight inclination

N = Deviation angle (n_1+n_2) with slopes - n_1 and $+n_2$

Substituting for a = 1°, h_1 = 0.75 m,

$$L = \frac{NS^2}{(1.5 + 0.035S)}$$

b. L < SSD

Let the vehicle be at the start of the valley curve or at the tangent point, TP, for minimum sight distance. Then,

$$h_1 + S\tan\alpha = \left(S - \frac{L}{2}\right)N$$

Therefore, $L = 2S - \dfrac{(2h_1 + 2S\tan\alpha)}{N}$

Substituting for a = 1°, h_1 = 0.75 m in the above equation gives,

$$L = 2S - \frac{(1.5 + 0.35S)}{N}$$

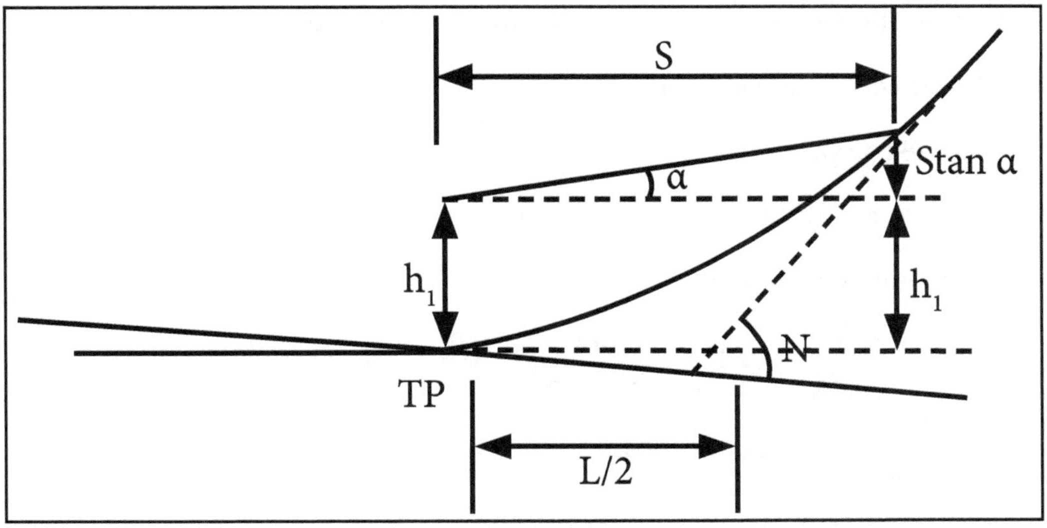

Headlights right distance when L < S.

To provide the cross drainage facility, the lowest point on the valley curve should be located. If the gradients are equal, then the lowest point will lie on the bisector of the angle between the curves. On unequal grades, the lowest point lies on the side of the flatter gradient.

Problems

1. A descending gradient of 1 in 25 meets an ascending gradient of 1 in 45 on a district highway in plain terrain. Let us compute the length of the sag curve which should provide safe driving at night.

Solution:

Given:

$$\text{descending gradient, } -n_2 = \frac{1}{25}$$

$$\text{ascending gradient, } n_1 = \frac{1}{45}$$

Formula to be used:

$$L = \frac{NS^2}{(1.5 + 0.035S)}$$

$$N = \frac{1}{35} - \left(-\frac{1}{45}\right) = 0.04 + 0.022$$
$$= 0.060$$

For a District Highway in plain terrain, the stopping distance = 120 m.

i. Length of sag curve = 0.38 $(NV^3)^{1/2}$

= 0.38 $(0.062 \times 80^3)^{1/2}$

= 67.70 m.

ii. Assuming L > S, then

$$L = \frac{NS^2}{1.5 + 0.035S} = \frac{0.062 \times 120^2}{1.5 + 0.035 \times 120}$$
$$= \frac{892.8}{5.7} = 156.63\,m = 160\,m.$$

As this value is greater than stopping distance of 120m, this assumption is correct.

Hence, L = 160m may be provided which satisfies the headlight requirements also.

Hence, the length of valley curve may be taken as 132 m.

2. A road with an ascending gradient of 1 in 120 is followed by a descending gradient of 1 in 100 and the design speed on the road is 100kmph and the overtaking sight distance is 510 m. Let us calculate the length of the summit curve.

Solution:

Given:

Ascending gradient = 1 in 120

Descending gradient = 1 in 100

Design speed on the road = 100kmph

Overtaking sight distance = 510 m

Therefore, $N = \frac{1}{120} - \left(-\frac{1}{100}\right) = \frac{11}{600}$

Formula to be used:

As this value is less than OSD of 510 m, assume L is less than OSD.

$$\begin{aligned} L &= 2S - \frac{9.6}{N} \\ &= 2 \times 510 - \frac{9.6 \times 600}{11} \\ &= 1020 - 523.6 \\ &= 496.4\,\text{m} \end{aligned}$$

Then for L < OSD,

this value is less than the OSD of 510 m.

Length of summit curve = 496.4 m

3

Traffic Engineering

3.1 Basic Parameters of Traffic: Volume, Speed and Density

The traffic engineering pertains to the analysis of the behavior of traffic and to design the facilities for a smooth, safe and economical operation of the traffic. Traffic flow, like the flow of water, has several parameters associated with it. The traffic stream parameters provides with the information regarding the nature of traffic flow, which helps the analyst in detecting any variation in the flow characteristics. Understanding traffic behavior requires a thorough knowledge of traffic stream parameters and their mutual relationships.

3.1.1 Traffic Volume Studies

There are two ways of counting the number of vehicles on a road practically. One is flow or volume, which is defined as the number of vehicles that pass a point on a line or a highway or direction of a highway during a specific time interval.

The measurement is done by counting number of the vehicles, n_t, passing a particular point in one lane in a defined period t. Then flow q expressed in vehicles/hour and is given by,

$$q = \frac{n_t}{t}$$

Flow is expressed in planning and the design field taking a day as the measurement of time.

Variations of Volume

Variation of the volume with time, i.e. month to month, day to day, hour to hour and within an hour is also as important as the volume calculation. Volume variations can also be observed from season to season.

In a pleasant motoring month of summer, the volume will be above average but will be more pronounced in the rural than in urban area. But this is the most consistent of all variations and affects the traffic stream characteristics least.

Weekdays, Saturdays and Sundays will also face a difference in pattern. But comparing day with day, the patterns for routes of a similar nature often show a marked similarity, that is useful in enabling the predictions to be made.

The most significant variation is from hour to hour. The peak hour detected during mornings and evenings of weekdays, which is generally around 8 to 10 per cent of the total daily flow or 2 to 3 times average hourly volume. These trips are mainly the work trips, which are relatively stable with time and more or less constant from the day to day.

Types of Volume Measurements

Since there is considerable variation in the volume of traffic, various types of measurements of volume are commonly adopted which will average these variations into a single volume count to be used in several design purposes.

Average Annual Weekday Traffic (AAWT)

The average 24-hour traffic volume occurring on the weekdays over a full year. It is computed by dividing the total weekday traffic volume for the year by 260.

Average Annual Daily Traffic (AADT)

The average 24-hour traffic volume at a given location over a full 365-day year, i.e. the total number of vehicles passing the site in a year divided by 365.

Average Weekday Traffic (AWT)

An average 24-hour traffic volume occurring on the weekdays for some period of time less than one year, such as for a month or a season.

Average Daily Traffic (ADT)

An average of 24-hour traffic volume at a given location for some period of time less than a year. It can be measured for six months, a season, a month, a week, or as little as two days. An ADT is a valid number only for the period over which it was measured.

The relationship between AAWT and AWT is analogous to that between AADT and ADT. Volume in general is measured using several ways such as manual counting, detector/sensor counting, moving-car observer method, etc.

The volume study mainly establishes the importance of a particular route with respect to the other routes, distribution of the traffic on road, and the fluctuations in flow. All which eventually determines the design of a highway and related facilities. Thus, volume is treated as the most important of all the parameters of traffic stream.

3.1.2 Speed Studies

Speed is considered as a quality measurement of travel as the drivers and passengers will be concerned more about the speed of the journey than the design aspects of the traffic. It is defined as the rate of the motion in distance per unit of time.

Mathematically speed or velocity v is given by,

$$v = \frac{d}{t}$$

where,

v is the speed of the vehicle in m/s.

d is the distance traveled in m in time t seconds.

Speed of different vehicles will vary with respect to time and space. To represent these variation, various types of speed can be defined. Important among them are spot speed, journey speed, time mean speed, running speed, and space mean speed.

3.1.3 Speed and Delay Studies

Spot Speed

Spot speed is the instantaneous speed of a vehicle at a specified location. Spot speed can be used to design the geometry of road like horizontal and vertical curves, superelevation etc. Location and size of signs, safe speed, design of signals, and speed zone determination require the spot speed data.

Accident analysis, congestion and road maintenance are the modern fields of traffic engineer, which uses spot speed data as the basic input.

Spot speed can be determined using an endoscope, pressure contact tubes or direct timing procedure or radar speedometer or by the time-lapse photographic methods. It can be determined by speeds extracted from video images by recording the distance traveling by all the vehicles between a particular pair of frames.

Running Speed

Running speed is the average speed maintained over a particular course while vehicle is moving and is found by dividing the length of the course by time duration the vehicle was in motion. I.e., this speed does not consider the time during which the vehicle is made to stop, or has to wait till it has a clear road ahead.

$$V_e = \frac{\text{travelled distance}}{\text{time spent to cover the distance}}$$

The running speed will always be more than or equal to the journey speed, as the delays are not considered in calculating the running speed.

Journey Speed

Journey speed is the effective speed of vehicle on a journey between two points and is the distance between two points divided by the total time taken for the vehicle to complete the journey including any stopped time.

$$\frac{\text{travelled distance}}{\text{total time spent to cover the distance}}$$

If the journey speed is less than running speed, it indicates that the journey follows a stop-go condition with the enforced acceleration and deceleration.

The spot speed here may vary from zero to some maximum in excess of the running speed. A uniformity between journey and running speeds denotes comfortable travel conditions.

Time Mean Speed and Space Mean Speed

Time mean speed is defined as average speed of all the vehicles passing a point on a highway over some specified time period. The space mean speed is defined as the average speed of all the vehicles occupying a given section of a highway over some specified time period. Both mean speeds will always be different from each other except in unlikely event that all vehicles are traveling at the same speed.

Time mean speed is a point measurement while space mean speed is a measure relating to the length of highway or lane, i.e. the mean speed of vehicles over a period of time at a point in space is the time mean speed and the mean speed over a space at a given instant is the space mean speed.

$$V_r = \frac{\text{distance}}{\text{average of time spent to cover the distance}}$$

Delay Measures

Delay measures are used for freeways and signalized intersections to evaluate the benefits of operational improvements and to estimate cost-effectiveness.

Freeway delay occurs when travel speeds are less than some arbitrary "free-flow" threshold, generally 35 or 45 mi/h (56.3 or 72.4 km/h). The delay is measured in terms of flow rate and travel time in excess of free-flow value. Delay is expressed in vehicle-hours.

Speed and travel time data were historically recorded on "tachograph" charts by equipment installed in floating car vehicles and were reduced manually. The newer methods record and store the data in electronic form, allowing them to be downloaded into a personal computer for making the analysis easier.

For example, a surveillance system containing speed and traffic flow detectors spaced at 1/3-m (1800-f [548.6-m]) increments or less, depending on the desired accuracy of the delay estimate, can be used to construct the speed profiles.

With computer assistance, vehicle hours of delay can be calculated from the profiles without deploying the floating vehicles. The automatic data collection approach also makes it easier to gather data samples at more frequent time intervals during a day or on more days during a week.

Intersection delay can be characterized by the stopped delay, time in queue delay, and approach delay. Figure below depicts the time-space trace of a vehicle that comes to a stop several times on the same signalized intersection approach. This trace might occur at a congested location during the peak volumes when a queue of stopped vehicles is not completely discharged during one green phase. It also constitute a situation where there is considerable compression of the queue during a red phase, or where a lane is carrying both through and left-turn movements, or stop-and-go conditions exist.

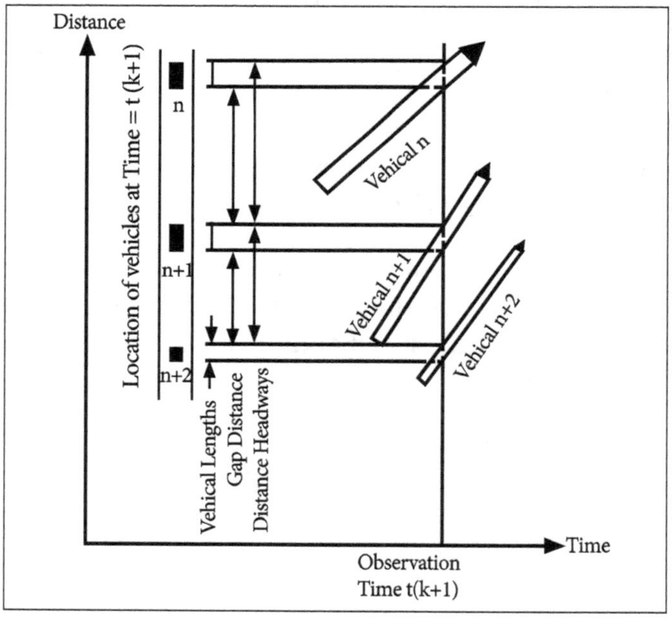

Distance headways of the consecutive vehicles at a particular time.

Stopped Delay

The sum of the stopped times, shown in the below figure, equals the total stopped delay. Stopped delay, an obvious measure of traffic to the motorist, is also important when evaluating the environmental impacts like the vehicle emissions.

Comparisons of stopped delay between intersections may be misleading when one intersection is operating under conditions where the time of stop is short, but low speed movement in a long sluggish queue follows stop.

An associated parameter, percent of vehicles stopping, while not strictly a delay measure, is a useful statistic, particularly in evaluating both fuel consumption and emissions. It is defined as number of vehicles that stop at least once, divided by the total number of vehicles using the approach and is expressed as a percent.

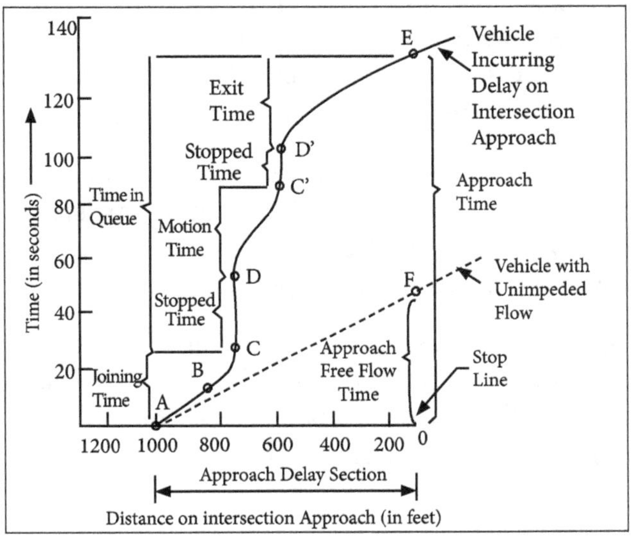

Time-space trace for a vehicle with multiple stops.

Time in Queue Delay

The time in queue is the sum of stopped time, motion time, and exit time as shown in the above figure. It is equivalent to the total approach time minus the joining time. The time in the queue delay is set equal to the time in queue, even though some progress is being made across the intersection during this time.

Time in queue delay often falls between stopped delay and approach delay. Time in queue delay can also be greater than approach delay because approach delay is equal to the difference between the actual time on the approach and the free-flow time of an unimpeded vehicle over the same distance.

Approach Delay

The approach time is the total time required to traverse the approach section under study. The approach delay is equal to approach time less the time required for an unimpeded vehicle to travel through the same section under free-flow conditions.

Approach delay may be used directly in the analysis of road-user time costs. This measure also compares the efficiency of the intersections having different modes of control, such as stop or yield signs and pre-timed or the vehicle-actuated traffic signals.

Delay data collection techniques have been primarily manual. However, the advent of areawide traffic control and system collected detector and signal status data have increased the information available for estimating delay at the signalized intersections. Data collection techniques include point sample, input-output, and path trace.

The point sample method periodically samples the intersection approach to record the number of stopped vehicles at the equally spaced time intervals. It determines the average number of stopped vehicles, calculates the total stopped time by multiplying the average number of stopped vehicles by the duration of the sample interval, and then calculates the average stopped delay by dividing total stopped time by the number of vehicles passing the study section. It is analogous to taking a series of snapshots at regular intervals.

The input-output method samples data during intervals, rather than between the intervals. It addresses the flow rates at the upstream and downstream boundaries of the approach area. The SCOOT adaptive traffic control system and the UTCS-enhanced areawide traffic control system use input-output technique.

The vehicle path trace method is essentially the same as the two other data collection techniques. The sample data are obtained by either running test vehicles through the approach area, measuring the various passage times of selected sample vehicles, or obtaining trace data from detector imaging techniques.

3.2 Parking Studies

Any vehicle will be parked at short time or longer time, provision of parking facilities is essential. Need for parking spaces is usually very great in areas of business, residential or commercial activities.

Park-and-ride

Providing adequate parking space to meet the demand for parking in the Central Business District (CBD). This problem usually confronts a city traffic engineer. Solution is not simple. Parking studies are used to determine the demand for and the supply of parking facilities.

Types of Parking Facilities

1. On-Street Parking Facilities

On-Street parking facilities are also known as curb facilities. Parking bays are provided alongside the curb on one or both sides of the street. Some of them are as follows:

i. Unrestricted parking.

ii. Unlimited and free.

iii. Restricted parking facilities.

They are limited to specific times for a maximum duration. It may or may not be free.

- Handicapped parking
- Bus stops
- Loading bays

2. Off-Street Parking Facilities

- Privately or publicly owned surface lots and garages
- Self-parking garages
- Attendant-parking garages

Parking Terms

A space-hour is a unit of parking that defines the use of a single parking space for a period of 1 hour.

Parking accumulation is the number of parked vehicles in a study area at any specified time.

Parking volume is the total number of vehicles that park in a study area during a specific length of time, usually a day.

Parking duration length of time a vehicle is parked at a parking indication of how frequently a parking space becomes available.

Parking load is the number of space-hours used during the specified period of time.

Parking turnover is the rate of use of a parking space. It is btained by dividing the parking volume for a specified period by the number of parking spaces.

Methodology of Parking Studies

1. Inventory of Existing Parking Facilities

It is the detailed listing of the location and all other relevant characteristics of each legal parking facility, private and public.

The study area includes both on-street and off-street facilities such as:

- Times of operation and limit on duration of parking, if any
- Type and number of parking spaces at each parking facility
- Parking fees, method of collection

- Type of ownership
- Restrictions
- Other restrictions, loading and unloading zones, bus stops, taxi ranks
- Permanency

The inventory should be updated at regular intervals of about four to five years.

2. Collection of Parking Data

Accumulation

By checking the amount of parking during regular intervals on different days of the week. It is carried out on an hourly or 2-hour basis. It is used to determine hourly variations of parking and peak periods of parking demand.

Turnover and Duration

Collecting data on a sample of parking spaces in a given block. Recording the license plate of the vehicle parked on each parking space in the sample at the ends of fixed intervals during study period. The length of the fixed intervals depends on the maximum permissible duration.

For example, if maximum permissible duration of parking at a curb face is 1 hour, a suitable interval is every 20 minutes.

If the permissible duration is 2 hours, checking every 30 minutes would be appropriate. Turnover is then obtained from the equation,

$$T = \frac{\text{number of distance vehicle parked}}{\text{number of parking spaces}}$$

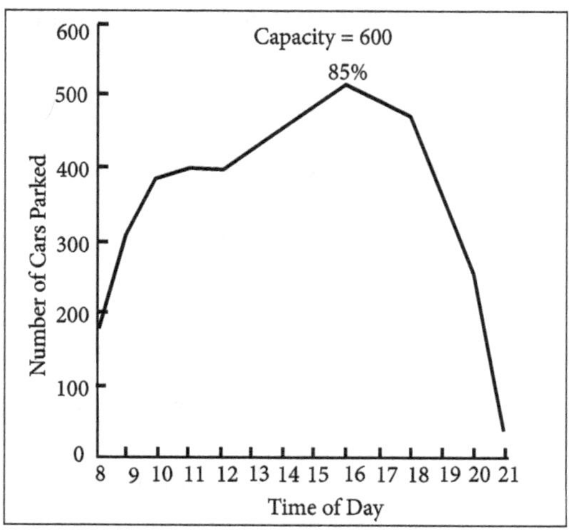

Parking accumulation at a parking lot.

Parking Demand

By interviewing drivers at the various parking facilities. Interview all drivers using the parking facilities on a typical weekday between 8:00 a.m. and 10:00 p.m.

Information includes (1) the trip origin, (2) purpose of trip, (3) driver's destination after parking.

The location of the parking facility, the times of arrival and departure, vehicle type. Parking interviews also can be carried out using the postcard technique about 30 to 50 percent of the cards distributed are returned.

Analysis of Parking Data

- Number and duration for vehicles illegally parked.
- Number and duration for vehicles legally parked.
- Supply of parking facilities.
- Space-hours of demand for parking.

The space-hours of demand for parking are obtained from the expression,

$$D = \sum_{i=1}^{N} (n_i t_i)$$

Where,

N = Number of classes of parking duration ranges

D = Space vehicle-hours demand for a specific period of time

n_i = Number of vehicles parked for the ith duration range

t_i = Midparking duration of the ith class

The space-hours of supply are obtained from the expression,

$$S = f\sum_{i=1}^{N} (t_i)$$

where,

N = Number of parking spaces available

S = Practical number of space-hours of supply for a specific period of time

t_i = Total length of time in hours when the ith space can be legally parked on during the specific period

f = Efficiency factor

3.3 Road Accidents: Causes and Preventive Measures

Road accidents is a global tragedy with ever raising trend. They are common fare in our daily news as we see these accidents happen almost everyday. Accidents are usually classified as single vehicle either colliding with fixed objects or multiple vehicle accidents in which two or more vehicle may collide head-on. Most people continue to be negligent and ignore the dangers involved in their driving and hence, these accidents happen.

Causes

1. Carelessness is one of the cause of road accidents in our country. Under carelessness, we have various examples which include, using a handheld mobile telephone while the vehicle is moving,driving through the red light, emerging from a side road into the path of another vehicle.

2. Lack of experience and licencing untrained drivers and also those with the blur eye sight who drive at night is one of the major cause of escalating accidents rate in this country.

3. Drunkeness: Most drivers drive under the influence of alcohol,intoxicated substances and under stress caused by economic or family problems which because of the state of the mind affect them causing road accidents. According to Dr. Lawrence Edusei, A Pathologist, from Korle-bu Teaching Hospital,In the year 2006, most of the drivers had high percentages of blood alcohol, which is far above the accepted level of 2.43/100ml. This is enough to cause the motor accidents.

4. Lack of sensitivity and responsibility on the part of state authorities to look into the situations on the roads like mal-functioning traffic lights. It shows that human sensibility and life respecting emotions of the state authorities have been dulled and desentilized on our roads. It is a sad reflection of deterioration in effectiveness and debility of governance in the country.

5. Disregarding of road signs and warning signs: Some drivers are more concerned about getting to their destination than of how they would get there. These drivers violate road signs and laws just to get to where they are going.

Preventive Measures

Issues that need to be considered in order to curb the accidents are policies made by the government to make sure that roads are well constructed and maintained.

There should be strategies that will eliminate the corruption at the D.V.L.A to make sure right thing is done. That's issuing of fake driving licence and also licence to individuals who have not gone through driving test by the D.V.L.A.

Passengers should also be vigilant enough to check drivers when they are overspeeding and driving under the influence of alcohol. Police men and women should help in playing major role to check drivers do the right thing to eliminate the road accidents.

Enforcements of appropriate legislation to make sure that all commercial and private drivers are well trained before they attain driving license to drive on the streets.

The Police should also make sure that all the broken down vehicles are towed off the streets.

The urgency and the need for action lies in the fact that the nation's best resources are its people. When road accidents are taken cognisance of, the cost of police and fire personnels responding to accidents will be reduced and productive time will be saved.

3.3.1 Condition Diagram and Collision Diagrams

Condition Diagram

Although a collision diagram is more prevalent, analysts often use condition diagrams to supplement the analysis before generating countermeasure ideas. Unlike collision diagrams, condition diagrams are drawn to scale providing the information on intersection (or segment) layout, roadway and lane widths, sight obstructions, crosswalks, traffic control devices, fixed objects, and other potentially noteworthy features. Condition diagrams can start as rough sketches in the field or be started in the office using aerial imagery, later filling in the details while in the held.

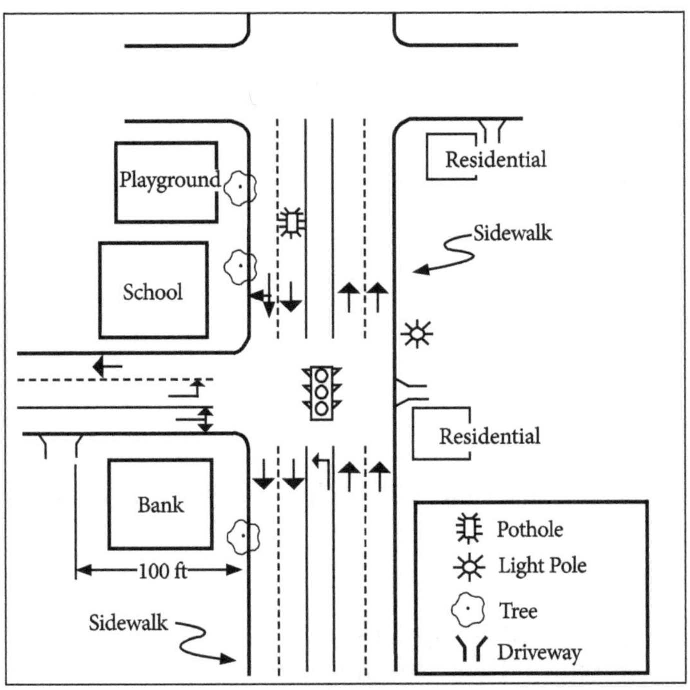

Example condition diagram.

Collision Diagram

Collision diagrams are used to display and identify similar accident patterns. They provide information on the type and number of accidents; including conditions such as time of day, day of week, pavement conditions, climatic conditions, and other information critical to determining the causes of safety problems.

Accident reports should be organized by year of occurrence and accident type for the analysis period. Accidents that occurred after the significant changes in highway or local land use should not be included.

A. Construction of Collision Diagram

Sketch the location diagram for either an intersection or roadway section using a standard form.

1. Sketch need not be to scale.
2. Show the path of each vehicle involved in the accident with adequate room for information.
3. Place a north arrow for orientation and any other descriptive information:
    a. Period of analysis
    b. Location identification
    c. Label the roads
4. Sketch the path of each vehicle to show vehicle maneuver, type of collision and accident severity, and:
    a. Time of day
    b. Date
    c. Day of week
    d. Pavement condition
    e. Lighting condition
    f. Other important information, such as alcohol involvement.

B. Symbols

Symbols representing the nature of operation, vehicle or the object involved and severity of the accident are adopted. These are shown in the example collision diagram on the following page. Symbols to represent types of collisions diagrams are also standardized.

C. Accident Patterns

Accident patterns seen on collision diagram can often suggest possible accident problems. For example, a large number of angle accidents would imply the potential of a sight distance restriction. At the end of this set of notes is a set of Accident Pattern table which identify accident type, possible causes, possible studies to determine the problem and possible counter measures.

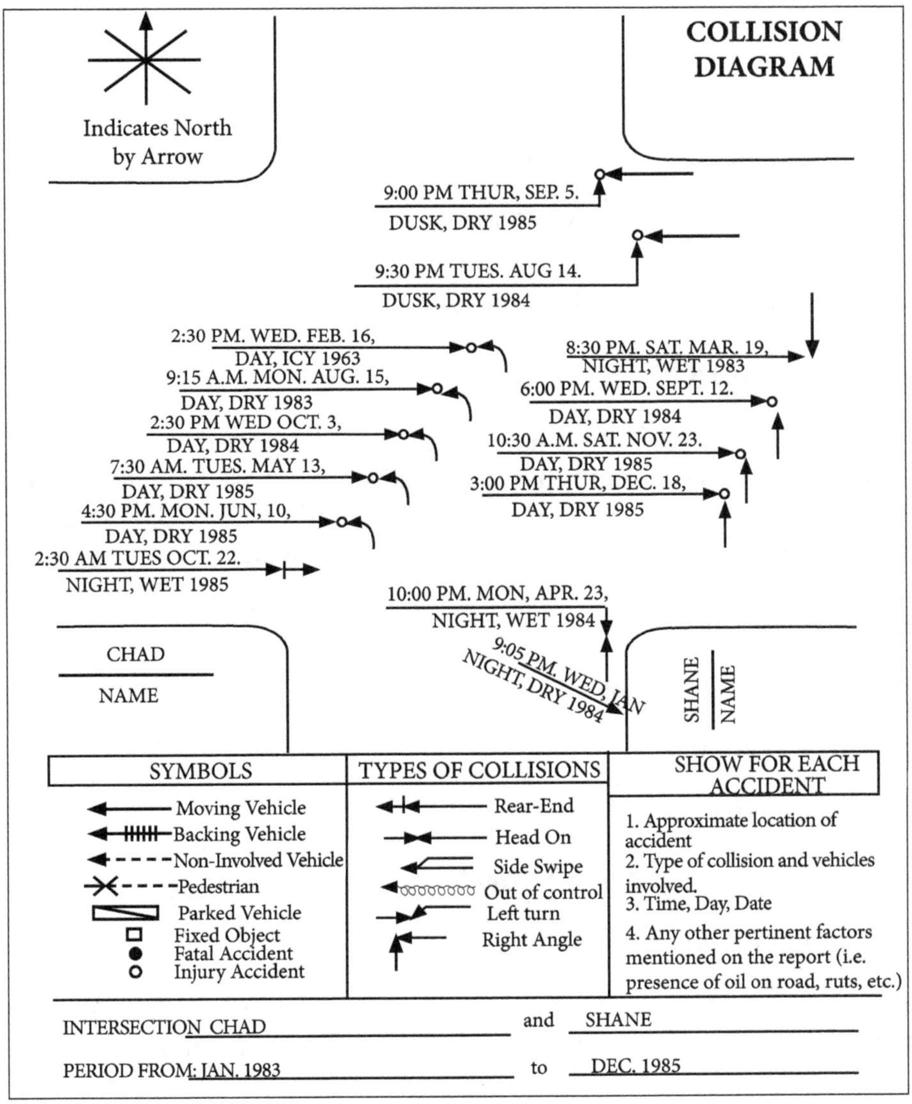

Example collision diagram.

3.3.2 PCU Factors

Passenger Car Unit (PCU) is a metric used in Transportation Engineering, to assess traffic-flow rate on a highway. A Passenger Car Unit is a measure of the impact that a mode of transport has on traffic variables (such as headway, speed, density) compared to a single standard passenger car. This is also known as passenger car equivalent. For example, typical values of PCU (or PCE) are:

Table: Values of PCU.

Motorcycle	0.5
Car	1.0
LCV	2.2
Bicycle	0.2
3-wheeler	0.8
Bus, Truck	3.5

Highway capacity is measured in PCU/hour daily.

Determination of PCU

Traffic in many parts of the world is heterogeneous, where road space is shared among many traffic modes with different physical dimensions. Loose lane discipline prevails; car following is not the norm. This complicates computing of PCU. Some of the methods for determining passenger car units (PCU) are following:

- Chandra's Method
- Modified Density Method
- Headway Method
- Method based on Relative Delay
- Multiple Linear Regression Method
- Simulation Method

It may be appropriate to use different values for the same vehicle type according to circumstances like volume of traffic, speed of vehicle, lane width and several external factors.

3.3.3 Capacity of Highways

It is the ability of a road to accommodate traffic volume. It is the maximum hourly rate at which vehicles can reasonably be expected to cross a point on a roadway during a given time period under prevailing traffic roadway and control condition. For multi lane highway it is 2000pcphpl, and for 2-lane highway is 2800pcph.

Units: vphpl (vehicle Per hour per lane) or vph (vehicle per hour).

Types of Capacity

Basic capacity

The maximum volume of vehicles per hour that may pass a certain point or section of a road in a given time under the ideal condition.

It assume that all the vehicles are traveling at the same speed and minimum spacing is allowed.

Possible Capacity

The maximum number of vehicles that may pass a given section during a given period of time under prevailing (most frequent/usual) roadway, traffic and control condition.

Practical / Design Capacity

Capacity without the traffic density being so great as to cause unreasonable delays, restriction or hazard to the drivers freedom under the prevailing condition of roadway, traffic and control.

Basic capacity > possible capacity > design capacity

Factors Affecting Highway Capacity

1. Width of shoulder
2. Lane width
3. Commercial vehicles
4. Lateral clearance
5. Existence of intersections
6. Road alignment and geometry (curves, Superelevation etc)
7. Drivers and vehicular characteristics
8. One way or two way traffic and number of lanes
9. Flow speed
10. Single type or mixed traffic
11. Parking

12. Weather condition
13. Presence of pedestrians

3.3.4 LOS Concepts

Level-of-Service(LOS) of a traffic facility is a concept introduced to relate the quality of traffic service to a given flow rate. Level-of-Service is introduced by HCM to denote the level of quality one can derive from a local under the various operation characteristics and traffic volume. HCM proposes LOS as a letter that designate a range of operating conditions on a particular type of facility.

Six LOS letters are defined by HCM, namely A, B, C, D, E, and F, where A denote best quality of service and F denote the worst. These definitions are based on Measures of Effectiveness(MoE) of that facility.

Typical measure of effectiveness include speed, travel time, density, delay etc. There would be an associated service volume for each of the LOS levels. A service volume or service flow rate is the maximum number of vehicles, passengers, or the like, which can be accommodated by a given facility or system under given conditions at a given LOS.

Capacity

Such a facility is considered as an ideal facility and for such facilities the following values can be taken as capacity.

A capacity of 1900 vehicles per hour per lane for a speed of 80kmph.

A capacity of 2000 vehicle per hour per lane for a speed of 115kmph.

A capacity of 2800 vehicle per hour for both direction at 100kmph.

The above values are not analytical or experimentally derived, but, statistically derived from the observed field values from a large number of such sections. It is possible to have a flow higher than this capacity measure, but not necessary.

Level of Service

The above capacity value drop due to various 'non-ideal condition' which includes the changes in speed or travel time, traffic interruptions or restriction etc. Accordingly HCM has defined various levels of services for the traffic facility. Assigning quality value is based on several user surveys capturing the perception of the drivers on the quality of the traffic under various operating condition. The figure below illustrates the quality of services or Level-of-Services (A to F) and the various operating conditions.

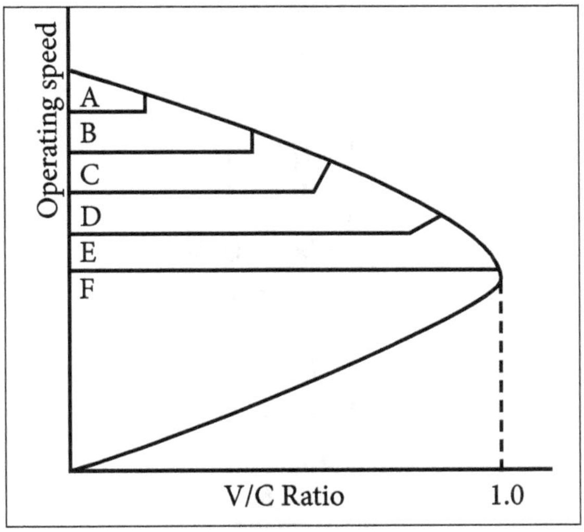

The LOS of a mid block section.

The same can be explained in the form of a table.

LOS	Quality	Speed (kmph)	V/C	Description
A	Free-flow	80	0.6	High level of physical and psychological comfort.
B	Reasonable free-flow	70	0.7	Reasonable level of physical and psychological comfort
C	Near free-flow	60	0.8	Local deterioration possible with blockages
D	Medium flow	50	0.85	Non-recoverable local disruptions
E	At capacity flow	40	0.9	Minor disturbances resulting breakdown
F	Congested flow	15	1.0	Break down of flow capacity drops

3.4 Road Traffic Signs

Traffic signs warn us of possible dangers and provides the information. They tell us what the rules are and what the road conditions are like.

Regulatory Signs

Regulatory signs tell us about the laws that must be obeyed.

Keep left sign.

Do not drive beyond this sign.

We must not turn right or make a U turn.

Don't make a U turn.

We must keep right.

This street now has lanes travelling in the opposite directions.

All traffic in the left lane must turn left.

We must turn right.

Except for STOP, GIVE WAY and ROUNDABOUT signs, most regulatory signs are rectangles. They are usually black on a white background. Sometimes they also have a colour such as red. Some parking signs are green on white.

Warning Signs

Warning signs tell you that there may be dangers ahead. They are usually black on a yellow background and are mostly diamond shaped.

Pictures, diagrams and symbols are used to alert you to danger.

Crossroads ahead

The road that we are travelling on ends, we must give way to all traffic.

Road conditions will change ahead to two lanes of oncoming traffic.

The road ahead curves to the right.

There is a sharp right turn in the road ahead.

Sharp bend to the left ahead.

Road divides ahead.

Divided road ends ahead.

Road narrows ahead.

A side road meets the road we are travelling on.

Winding road ahead.

Give way sign ahead.

Stop sign ahead.

Pedestrian crossing ahead.

Pedestrians may be crossing ahead.

Hospital ahead drive carefully.

Look out for bicycle riders.

Road is slippery when wet.

There is a hump in the road ahead.

Look out for kangaroos.

As we approach the hill ahead, we will not be able to see a safe distance in front of us. Drive carefully.

The road ahead is under water.

Grid ahead

A grid is a row of metal lengths across the road.

Water may cover the causeway ahead.

Beware of slow moving vehicles entering traffic.

Narrow bridge ahead, slow down and be prepared to give way or stop.

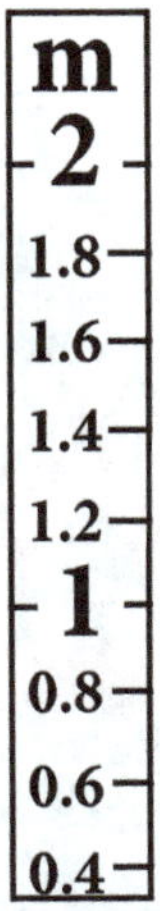

This sign shows the depth of floodwaters across the road.

This sign is used to advise drivers that the road ahead may be covered in floodwaters.

Advisory Signs

The advisory speed signs are sometimes used together with other signs. They show the maximum speed that is safe in good conditions. Sometimes the other signs are used together with warning signs to advise on how long we should look out for a particular hazard. Road bends to right. 55 km/h maximum speed in good conditions.

Stock Signs

When we see a sign with a picture of an animal, or words such as 'stock crossing', we may be approaching animals on or near the road. We must slow down or stop to avoid crashing with them. Fines apply if we do not obey these signs.

Stock may be crossing ahead.

Warning Triangles for Crashes and Breakdowns

Vehicles with a GVM over 12 tonnes must place atleast three warning triangles in breakdown situations.

Variable message signs.

Variable Message Signs are large electronic signs placed on side of some roads. They display messages to warn motorists of changes in normal traffic conditions on the road ahead, such as fog, crashes, road works, congestion and road closures. These early warnings help provide a safe and efficient traffic flow.

Image showing a variable message sign with the text "KEEP LEFT UNLESS OVERTAKING".

Motorway Signs

Motorway signs give information about the start and end of a motorway and the exits from a motorway.

These signs indicate the start of a motorway.	These signs indicate the end of a motorway.

Image showing two types of motorway exit signs:

Exit from motorway in 2km.	Exit from motorway now.

Priority for Buses

In a built-up area a bus that has a GIVE WAY sign displayed and signals its intention to pull out from its stopping place, has priority over other traffic, travelling in the left lane or left line of traffic.

Sign indicating that buses have priority over the other traffic.

Temporary Roadwork Signs

These signs are used at roadworks. Slow down, look out for any hazards and be prepared to stop. We must obey any signal from a traffic controller and give way to any worker.

The speed limit signs at the work zones are enforceable and must be obeyed.

Traffic controller ahead be prepared to stop (night).

Traffic controller ahead be prepared to stop.

Workers ahead (day).

Image showing two signs in which one when approaching roadworks and another when roadworks end:

Approaching roadworks, slow down and be prepared to stop.

End road work.

Stop and slow sign and prepare to stop.

These signs are held by traffic controllers. We must obey them. Two types of motorway signs. One for road hazard ahead and another for temporary traffic lights in place.

Image showing two signs: One for when there are no lines marked on road and the other for when there is a closed lane ahead.

No lines marked on road. Take care if overtaking.

Closed lane ahead.

One image warn the drivers for loose stones and the other indicating that there is a 40km/h speed limit for road work.

Watch out for loose stones.

40km/h speed limit for road work.

3.5 Road Markings

Road markings are used as a means of controlling and guiding traffic. They are highly important on urban roads and intersections as they promote road safety and bring out smooth and harmonious flow of traffic along guided paths of travel. They also supplement messages conveyed by road signals and signs. In some cases, they are used alone to convey certain regulation, information or the warning that cannot otherwise be effectively made known to the road users.

Road surface markings are the devices on a road surface in order to convey official information. Road surface markings are used on paved roadways to provide guidance and the information to drivers and pedestrians. Uniformity and standardisation of the markings is an important factor in minimizing confusion and uncertainty about their meaning. These do help in reducing the accidents and manpower requirements for regulating traffic.

Functions of Road Markings

The main functions of the road markings are to guide the safe and smooth flow of traffic in the following ways:

i. Stop and go

ii. Segregation of traffic

iii. Overtaking or not

iv. Give way instruction

v. Inter-vehicle distance

vi. Two lanes to one lane/ lane traffic

vii. Speed indication

viii. Parking zone or no parking

ix. One way

x. Direction

xi. Type of vehicles allowed

xii. Pedestrian crossing

Road markings play a useful role in traffic management. They should convey the required information to the driver without distracting his attention from the carriageway, for which the following rudimentary requirements should be met:

i. Night visibility i.e., retroflectivity

ii. Day visibility i.e., whiteness/contrast

iii. Anti-skid

iv. Wet night visibility (humid/rain/ fog)

3.6 Types of Intersections

Intersection is an area shared by two or more roads. This area is designated for the vehicles to turn to different directions to reach their desired destinations. This is because vehicles moving in different direction want to occupy same space at the same time

Types

There are two main types of intersection of roads such as:

1. Grade-separated intersections or interchanges Ø
2. At-grade intersections

Grade Separated Intersection

It is a bridge that eliminates crossing conflicts at intersections by vertical separation of roadways in the space. Route transfer at grade separations is accommodated by the interchange facilities consisting of ramps. The interchange configurations are designed in a way to accommodate economically the traffic requirements of flow, operation on the crossing facilities, physical requirements of the topography, adjoining land use, type of controls, right-of-way and direction of movements.

At-Grade Intersections

At-grade intersections is the intersection in which all the exchanges between the roads take place on the same plane. These are of two main types:

1. Standard at-grade intersections Ø
2. Round about at-grade intersections

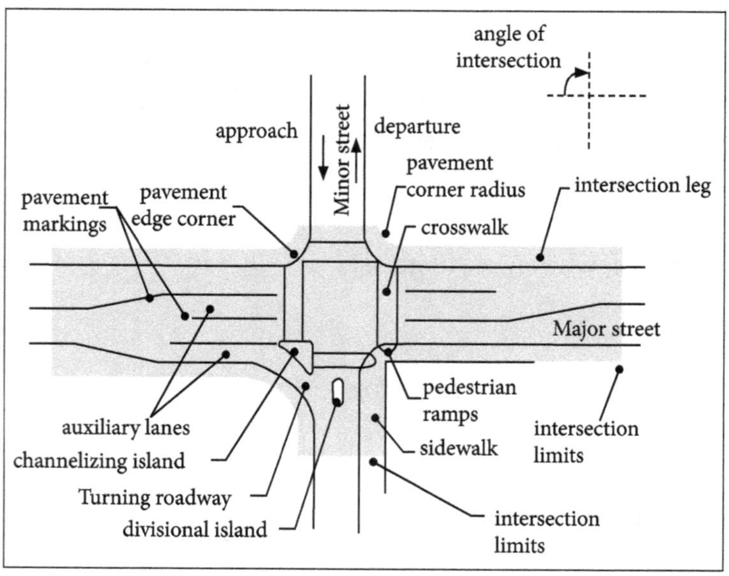

Key elements of intersection of road.

Different Types of At-Grade Intersections

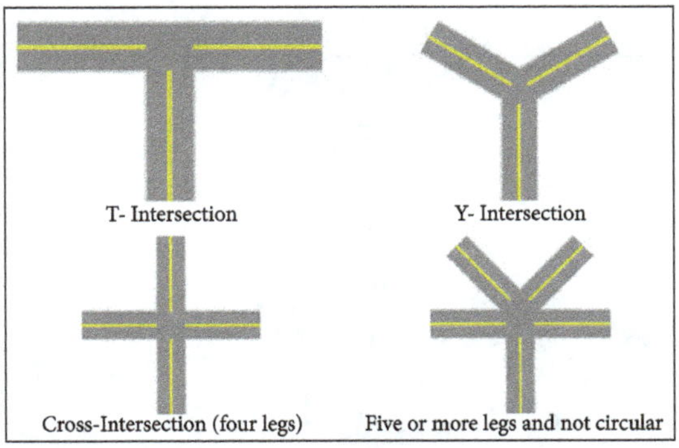

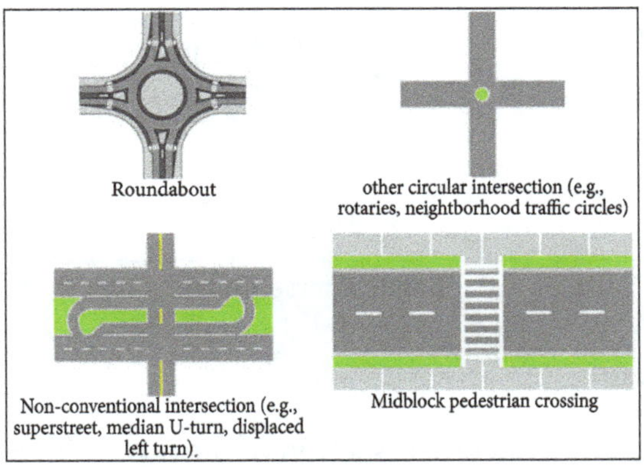

Most intersections have three or four legs, but multi-leg intersections are not unusual.

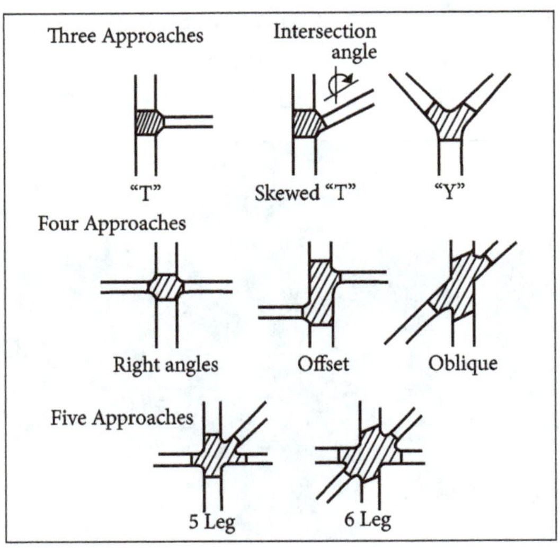

3.6.1 Design of Plain, Flared, Rotary and Channelized Intersections

Simple Intersections

Simple intersections maintain the street's typical crosssection and number of lanes throughout the intersection, on both major and minor streets. Simple intersections are best-suited to locations where auxiliary lanes are not needed to achieve the desired level-of-service.

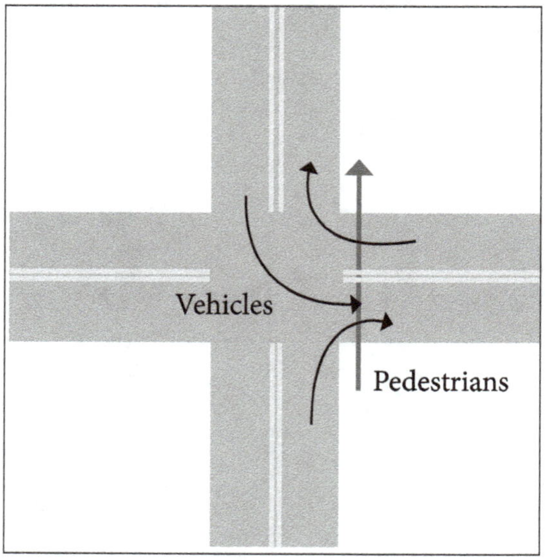

Simple intersections

Flared Intersections

Flared intersections expands the cross-section of the street. The flaring is often done to accommodate a left-turn lane, so that the left-turning bicycles and motor vehicles are removed from the through-traffic stream to increase capacity at high-volume locations, and safety on higher speed streets. Right-turn lanes, less frequently used than left-turn lanes, are generally a response to large volumes of right turns.

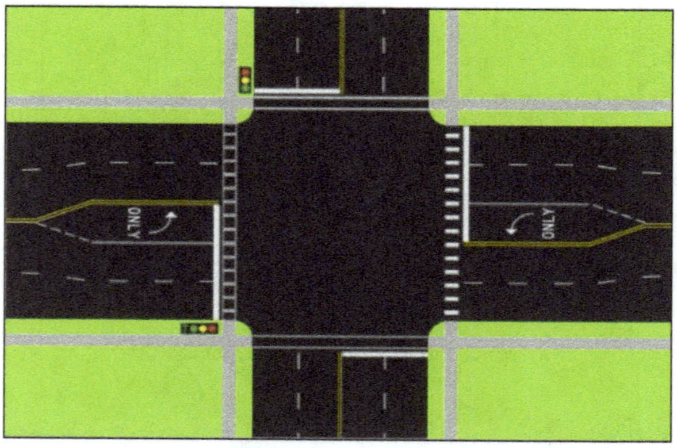

Flared intersections

Roundabout Intersections

A roundabout is a type of circular intersection or junction in which road traffic flows almost continuously in one direction around a central island. It gives maximum safety in all types of intersections.

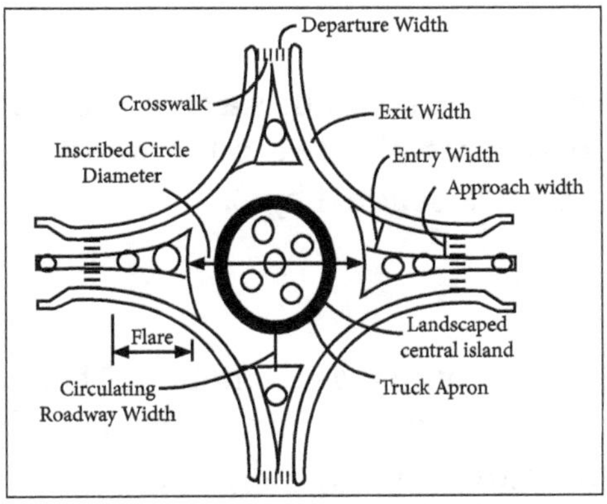

Roundabout intersections

Channelized Intersections

Vehicles approaching an intersection are directed to definite paths by islands, marking etc. and this method of control is known as channelization. Channelized intersection provides more safety and efficiency. It reduces the number of the possible conflicts by reducing the area of conflicts available in the carriageway.

If no channelizing is provided, the driver will have less tendency to reduce the speed while entering the intersection from the carriageway.

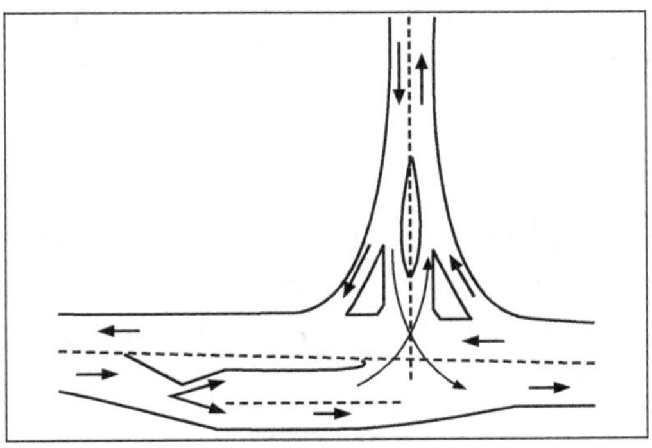

Channelized intersections

3.7 Design of Traffic Signals

Definitions and Notations

A number of definitions and notations need to be understood in signal design. They are discussed below:

Cycle

A signal cycle is one complete rotation through all the indications provided.

Cycle Length

Cycle length is the time in seconds that it takes a signal to complete one full cycle of indications. It indicates the time interval between the starting of green for one approach till the next time green starts. It is denoted by C.

Interval

Thus it indicates the change from one stage to another. There are two types of intervals, change interval and clearance interval. Change interval is also called the yellow time indicates interval between the green and red signal indications for an approach. Clearance interval is also called all red and is provided after each yellow interval indicating a period during which all signal faces show red and is used for clearing off vehicles in the intersection.

Green Interval

It is the green indication for a particular movement or set of movements and is denoted by G_i. This is the actual duration the green light of a traffic signal is turned on.

Red Interval

It is red indication for a particular movement or set of movements and is denoted by R_i. This is the actual duration the red light of a traffic signal is turned on.

Phase

A phase is the green interval plus the change and the clearance intervals that follow it. Thus, during green interval, non conflicting movements are assigned into each phase. It allows a set of the movements to flow and safely halt the flow before the phase of another set of movements start.

Lost Time

It indicates the time during which the intersection is not effectively utilized for any movement. For example, when the signal for an approach turns from red to green, the driver of the vehicle which is in front of the queue, will take some time to perceive the signal (usually called as reaction time) and some time will be lost before vehicle actually moves and gains speed.

Phase Design

The signal design procedure involves six major steps. It includes:

1. Phase design

2. Determination of amber time and clearance time
3. Determination of cycle length
4. Apportioning of green time
5. Pedestrian crossing requirements
6. Performance evaluation of the design obtained in the previous steps

Objective of the phase design is to separate the conflicting movements in an intersection into various phases, so that movements in a phase should have no conflicts. If all the movements are to be separated with no conflicts, then a large number of the phases are required. In such a situation, the objective is to design phases with minimum conflicts or with a less severe conflicts.

There is no precise methodology for the design of phases. This is often guided by the geometry of the intersection, the flow pattern especially the turning movements, and the relative magnitudes of flow. Therefore, trial and error procedure is often adopted. However, phase design is very important because it affects the further design steps.

Further, it is easier to change the cycle time and green time when flow pattern changes, where as a drastic change in flow pattern may cause considerable confusion to the drivers. To illustrate various phase plan options, evaluate a four-legged intersection with through traffic and right turns. Left turn is ignored.

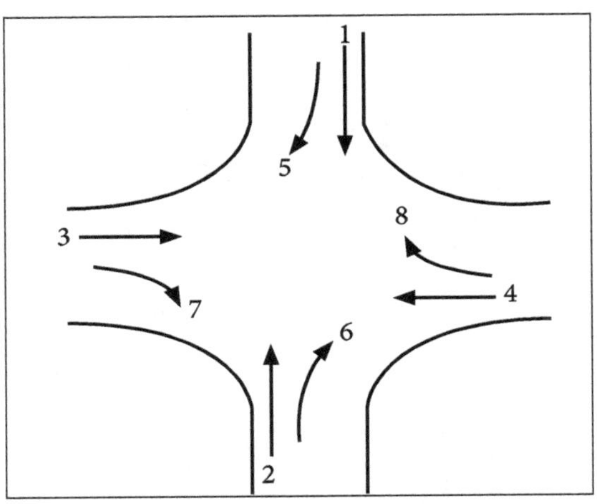

Four-legged intersection.

First the issue arise to decide how many phases are required. It is possible to have two, three, four or even more number of phases.

Two Phase Signals

Two phase system is usually adopted if through traffic is significant compared to the turning movements. For example, non-conflicting through traffic 3 and 4 are grouped in a single phase and the non-conflicting through traffic 1 and 2 are grouped in the second phase.

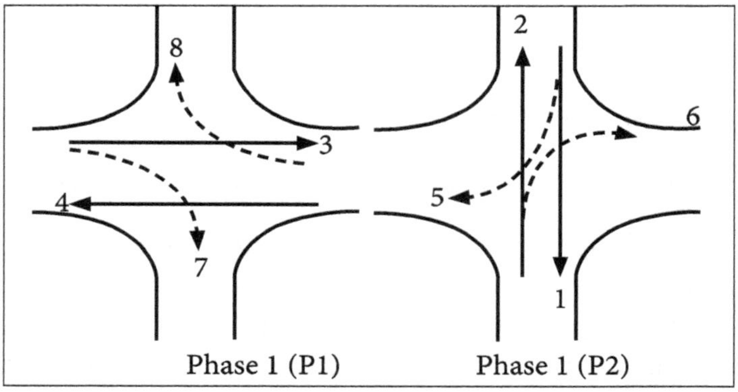

Movements in the two phase signal system.

However, in the first phase flow 7 and 8 offer some conflicts and are called permitted right turns. Needless to say that such a phasing is possible only if turning movements are relatively low. On the other hand, if turning movements are significant, then a four phase system is usually adopted.

Four Phase Signals

There are at least three possible phasing options. For example, the below figure shows the most simple and trivial phase plan.

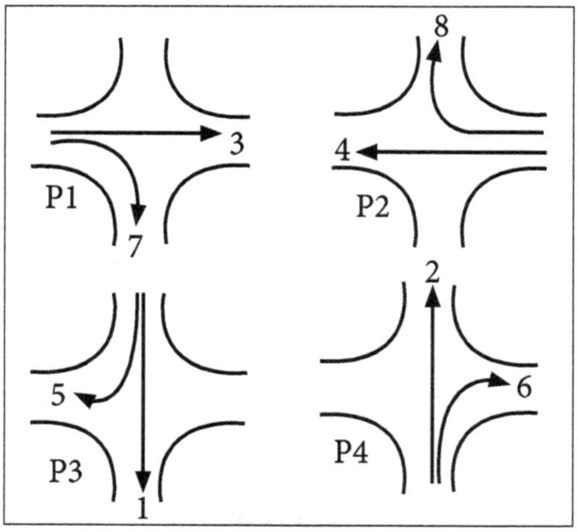

Movements in four phase signal system: option 1.

Here, the flow from each approach is put into a single phase avoiding all conflicts. This type of phase plan is ideally suited in urban areas where the turning movements are comparable with through movements and when through traffic and turning traffic need to share same lane. This phase plan can be very inefficient when turning movements are relatively low. Figure below shows a second possible phase plan option where opposing through the traffic are put into same phase.

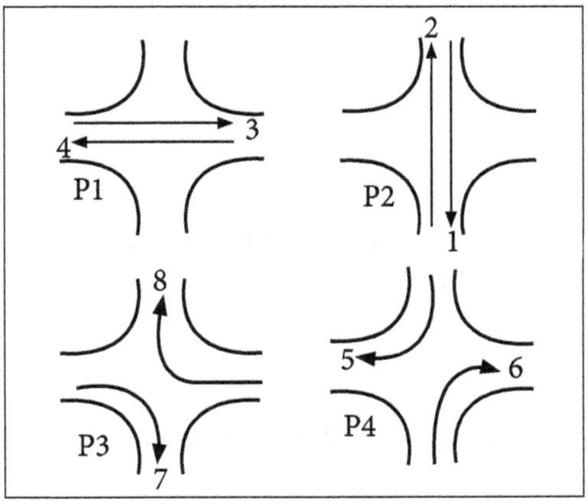

Movements in four phase signal system: option 2.

The non-conflicting right turn flows 7 and 8 are grouped into a third phase. Similarly flows 5 and 6 are grouped into the fourth phase. This type of phasing is very efficient when the intersection geometry permits to have at least one lane for each movement, and the through traffic volume is significantly high. Figure 5 shows yet another phase plan. However, it is rarely used in practice.

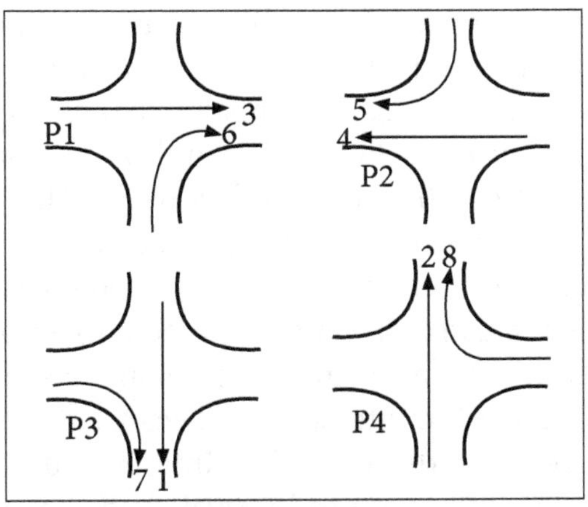

Movements in four phase signal system: option 3.

There are five phase signals, six phase signals etc. They are normally provided if the intersection control is adaptive, that is, signal phases and timing adapt to the real time traffic conditions.

3.7.1 Webster Method

Webster has shown that for a wide range of practical conditions minimum intersection delay is obtained when the cycle length is obtained by the equation.

$$C_o = \frac{1.5L + 5}{1 - \sum_{i=1}^{\phi} Y_i}$$

Where,

C_o = Optimum cycle length

L = Total lost time per cycle (sec)

Y_i = Maximum value of the ratios of approach flows to saturation flows for all lane groups using phase i (i.e., q_{ij}/S_j)

Φ = numbers of phases

q_{ij} = Flow on lane groups having the right of way during phase i

S_j = saturation flow on lane group j

This equation assumes that arrivals are random and traffic conditions at the intersection are undersaturated (i.e., demand flow is less than capacity).

Total Lost Time

Figure shows a graph of rate of discharge of vehicles at various times during a green phase of a signal cycle at an intersection. At the beginning of the green phase, the drivers of the first few vehicles in the queue take some additional time to react and accelerate and achieve their running speeds. This results in these vehicles initially traveling with headways that are higher than the saturation flow headway, thereby losing some time before the vehicles move with saturation headways. This lost time is the start-up lost time, designated as l_1. At the end of the green indication, there is a portion of the yellow indication that is always used by drivers to go through the intersection. This initial part is known as the extension of the effective green (e). However, sometime toward the end of the change period (yellow change interval and red clearance interval) is not used by vehicles to go through the intersection, because drivers are concerned about not complying with the yellow indication. The difference between the change period (yellow change interval and red clearance interval) and the extension of the effective green time (e) is designated as the clearance lost time (l_2). Suggested values for l_1 and l_2 are 2 seconds each. The total lost time (l_t) for each phase is the sum of l_1 and l_2 and can be expressed as,

$$l_t = l_1 + l_2 = l_1 + Y + R_c - e$$

where,

l_t = Phase lost time

l_1 = Start-up lost time = 2.0 sec

l_2 = Clearance lost time = Y + R_c - e (sec)

e = Extension of effective green time = 2.0 sec

Y = Yellow change interval (sec) and

R_c = Red clearance interval (sec)

Total lost time for a cycle is given as,

$$L = \sum_{1}^{\varphi} l_i$$

where Φ is the number of phase in a cycle.

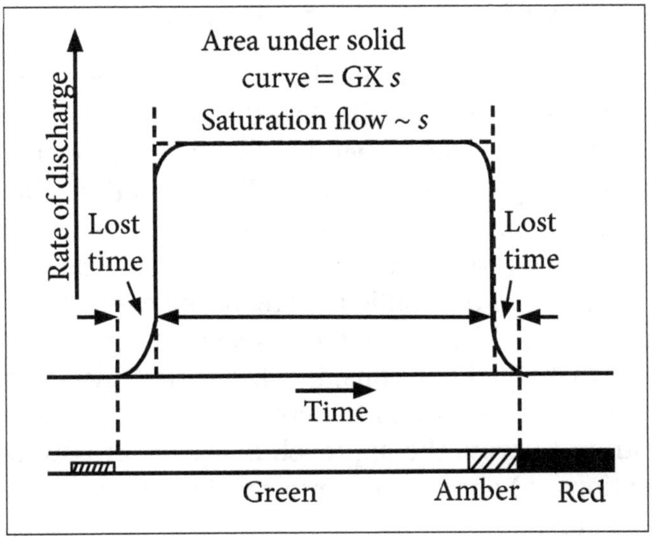

Discharge of vehicles at various times during a green phase.

Allocation of Green Times

In general, the total effective green time available per cycle is given by,

$$G_{te} = C - L = C - \left(\sum_{i=1}^{\phi} l_i\right)$$

where,

C = Actual cycle length used (usually obtained by rounding off C_o to the nearest 5 sec)

Gt_e = Total effective green time per cycle

To obtain minimum overall delay, the total effective green time should be distributed among the different phases in proportion to their Y values to obtain the effective green time for each phase.

$$G_{ei} = \frac{Y_i}{Y_1 + Y_2 \cdots Y_\phi} G_{te}$$

And the actual green time for each phase is obtained as,

$$G_{a1} = G_{e1} + l_1 - Y_1$$
$$G_{a2} = G_{e2} + l_2 - Y_2$$
$$G_{a3} = G_{e3} + l_3 - Y_3$$
$$G_{a\phi} = G_{e\phi} + l_\phi - Y_\phi$$

Minimum Green time for Pedestrian Movement

At an intersection where a significant number of pedestrians cross, it is necessary to provide a minimum green time that will allow the pedestrians to safely cross the intersection. Length of this minimum green time may be higher than that needed for vehicular traffic to go through the intersection. The green time allocated to the traffic moving in the north-south direction should, therefore, not be less than the green time required for pedestrians to cross the east-west approaches at the intersection. Similarly, the green time allocated to the traffic moving in the east-west direction cannot be less than that required for pedestrians to cross the north-south approaches. The HCM gives three equations that should be considered. The first shown as equation (1) considers the time taken by the pedestrian to perceive the signal indication and the time taken by the pedestrian to cross the approach along the crosswalk. The second and third shown as equations (2) and (3) take into consideration the time that is required for cyclic pedestrian demand.

$$G_{P,min} = t_{pr} + \frac{L_{cc}}{S_p} - y - R_c \qquad ...(1)$$

$$t_{ps} = 3.2 + \frac{L_{cc}}{S_p} + 9\frac{N_{ped}}{W_E} \quad \text{for } W_E > 3\,\text{m} \qquad ...(2)$$

$$t_{Ps} = 3.2 + \frac{L}{S_p} + \left(0.27 N_{ped}\right) \quad \text{for } W_E \leq 3\,\text{m} \qquad ...(3)$$

Where,

$G_{p,min}$ = Minimum green time based on pedestrian crossing time (sec)

t_{pr} = Time taken by the pedestrian to perceive the signal indication and depart the curb (HCM suggested value = 7 sec)

L_{cc}= Curb to curb crossing distance (m) or crosswalk length (m)

S_p = Average speed of pedestrians, usually taken as 1.0 m/sec (however, HCM 2010 suggests 1.2 m/sec when the percentage of elderly pedestrians is 20 percent or less and 1.0 m/sec when the percentage of elderly pedestrians is higher than 20 percent; also, an upgrade of 10 percent or more results in reduction of pedestrian speeds by 0.1 m/sec)

Y = Yellow change interval (sec)

R_c = Red clearance interval (sec)

W_E = Effective crosswalk width (m)

N_{ped} = Number pedestrians crossing during an interval = $N_{ped} = \frac{V_{ped,i}}{3600}C$

$V_{ped,i}$ = Pedestrian flow rate in the subject crossing for travel direction i (p/h)

Note that in some cases when left turns are permitted or protected-permitted, the reduction of the minimum green time by the yellow change interval (Y) and the red change interval (R_c) as shown in equation (1) result in pedestrians conflicting with left-turning vehicles. When it is feasible, the reduction of the minimum green by Y and R_c should be discarded and the minimum green is given as $G_{p,min} = t_{pr} + (L_{cc}/S_p)$.

3.7.2 IRC Method

- The pedestrian green time required for the major and minor roads are calculated based on walking speed of 1.2 m/sec. and initial walking time of 7.0 secs. These are the minimum green time required for the vehicular traffic on the minor and major roads respectively.
- The green time required for the vehicular traffic on the major road is increased in the proportion to the traffic on the two approach roads.
- The cycle time is determined after allowing amber time of 2.0 secs each.
- Minimum green time required for clearing vehicles arriving during a cycle is a determined for each lane of the approach road assuming that the first vehicle can take 6.0 secs. And the subsequent vehicles (PCU) of the queue will be cleared at a rate of 2.0 secs. The minimum green time required for the vehicular traffic on any of the approaches is limited to 16 secs.
- Optimum signal cycle time is calculated using Webster's formula The saturation flow values may be assumed as 1850,1890, 1950, 2250, 2550 and 2990 PCU per hour for approach roadway widths (keb to median or centerline) of 3.0, 3.5, 4.0, 4.5, 5.0 and 5.5m; for width above 5.5m, the saturation flow can be assumed as 525 PCU per hour per meter width. The lost time is estimated from the amber time, the inter-green time and the initial delay of 4.0 secs. for the first vehicle, on each leg.
- Signal cycle time and the phases can be revised keeping in view the green time required for clearing vehicles and the optimum cycle length determined in steps (iv) and (v) above.

4

Highway Materials

4.1 Subgrade Soil

Classification

Properties of subgrade soil are important for the design of pavement structure to withstand adverse weather conditions. Soil is also used in embankment construction which should provide adequate stability and also incompressible. It is an unconsolidated material that has resulted from the disintegration of rocks. It constitutes the earth's surface.

The type and characteristics of soil depend largely on its origin. Soil deposits constitute an assemblage of solid particles with formation of certain pore or void spaces. Voids are in turn filled with a gas or liquid or both. Three components viz., solid, gas and liquid play an appreciable role in the fundamental behaviour of soil.

Desirable Properties of Soils

The properties of a soil depend on size, shape, surface texture and the chemical composition of its grains. Density and moisture content influence to great extent the engineering properties of a soil.

Soils to be used as a highway material should have the following properties:

1. Short and long term stability of the subgrade and slopes of embankment
2. Compressibility within permissible limits
3. Adequate permeability
4. Compaction should be ease and economical
5. Minimum volume change at all weather conditions

4.1.1 Group Index

For qualitative evaluation of the desirability of a soil as a highway subgrade material, a number referred to as the group index has also been developed. The higher the value of the group index for a given soil, the weaker will be the soil's performance as a subgrade. A group index of 20 or more indicates a very poor subgrade material.

The group index GI can be expressed as,

$$GI = (F_{200} - 35)[0.2 + 0.005(LL - 40)] + 0.01(F_{200} - 15)(PI - 10)$$

where,

LL = Liquid limit

F_{200} = percent passing no. 200 sieve, expressed as a whole number

PI = Plasticity index

4.1.2 Subgrade Soil Strength

Tests on Soils

The suitability of soil is ascertained from various laboratory and field tests. Apart from the fundamental tests to determine the shear strength properties, permeability properties and settlement properties, the following laboratory and field tests are discussed.

1. Field Density Tests:
    a. Core-Cutter Method
    b. Sand Replacement Method
2. California Bearing Ratio Test:
    a. Laboratory Test
    b. Field Test

Plate Bearing Test

Field density of a soil is essential to assess the strength of a subgrade and shoulders. California Bearing Ratio test is an Adhoc test which gives an empirical strength number which is used to assess the bearing power of a subgrade. Plate Bearing test is used to find the modulus of subgrade reaction which is used in the pavement design.

Field Density Tests

Dry density of a soil in the field is called the field density or in-situ dry density or in-place dry density. Field density is found using adopting core-cutter method (IS: 2720 Part 29, 1975) or sand replacement method (IS: 2720 Part 20, 1974).

Core-cutter Method

1. Apparatus Required

i. Cylindrical core cutter of steel with steel doly and steel rammer

ii. Steel rule, palette knife, straight edge

iii. Balance of 1g sensitivity

iv. Apparatus for water content determination

v. Apparatus for extracting samples from the cutter

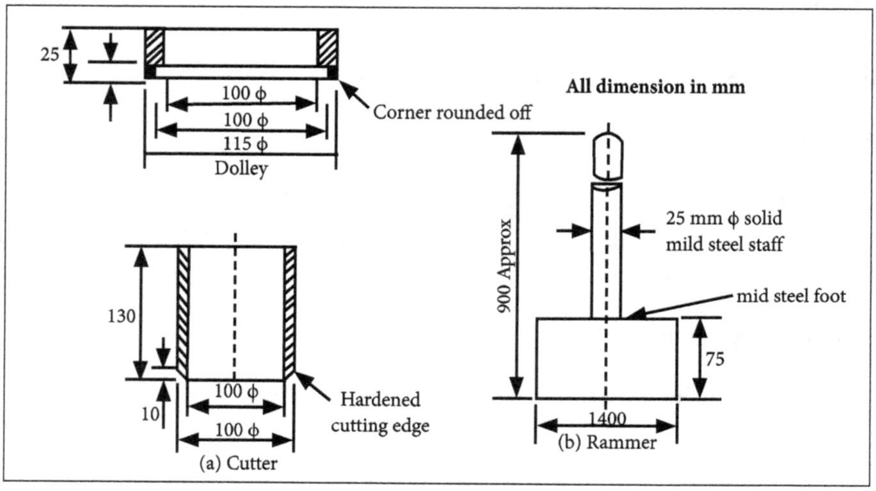

Apparatus for core cutter method (IS: 2720, Part 29, 1975).

2. Test Procedure

i. The inner dimensions of the core-cutter is measured (nearest to 0.25 mm) and its volume is found (V_c).

ii. The weight of the cutter is found without the dolly.

iii. A small area of about 300 mm² is cleared and levelled where the in-place density is to be determined.

iv. The cutter is placed on the levelled surface. The dolly is positioned on the cutter.

v. The cutter is advanced into the subsoil by the help of rammer until about 15 mm of the dolly protrudes above the surface.

vi. The soil around the cutter is dug, using a spade or pickaxe and the cutter is bodily removed with soil allowing some soil to project from the lower end of the cutter. The top and bottom of the cutter are trimmed by means of a palette knife and straight edge.

vii. The cutter with soil and without dolly is weighed (M_{sc}).

viii. The soil is removed from the cutter and the water content determined.

3. Computation and Result

Then Bulk density, $\rho = (M_{sc} - M_c)/V_c \;\; g/cc$

The Field density, $\rho_d = \rho/\left[1+(w/100)\right] \;\; g/cc$

where w is the percentage field water content.

The dry density and water content of soil are reported to the second decimal place and two significant figures respectively.

4. Applications and Limitations

i. The core cutter method is quite useful for fine-grained soils.

ii. It cannot be used on soils with more gravels or non-cohesive soils.

iii. It is a convenient and fast test.

As per IS norms repeat determinations should be done and average result should be reported.

This method is less accurate compared to sand replacement method.

Sand-Replacement Method

1. Apparatus Required

i. Sand pouring cylinder

ii. Cylindrical calibration can

iii. Clean and closely graded natural sand passing 1 mm IS sieve but retained on 600 µm IS sieve is cleaned

iv. Metal tray with hole - 300 mm², 40 mm deep, with a 100 mm hole in the center

v. Tools for excavating holes

vi. Glass plate - 450 mm², 9 mm thick

vii. Balance of 1 g sensitivity

viii. Apparatus for water content determination

ix. Metal tray or container

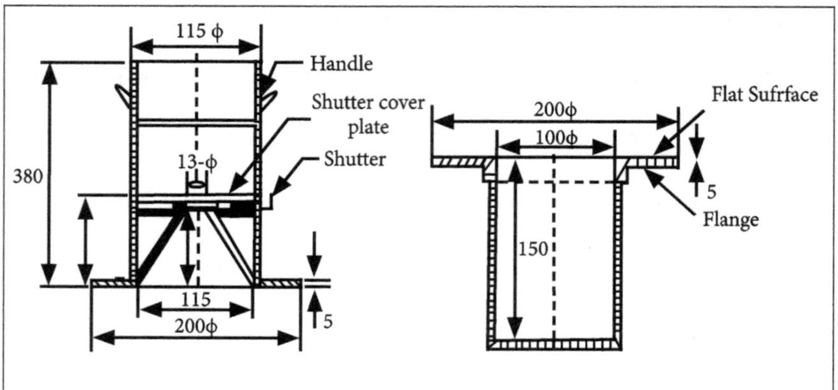

(a) Sand pouriing cylinder (b) Calibration can

Apparatus for sand replacement method.

2. Test Procedure

i. The pouring cylinder is filled with clean sand till the level of sand is about 10 mm from the top and the weight, M_1, is found.

ii. Mass is maintained constant throughout the test for which the pouring cylinder has to be calibrated.

iii. Pouring cylinder is placed on a glass plate and the top is closed when the conical portion has been filled.

iv. Sand on the glass plate is carefully collected and the sand is weighed. Steps i to iii are repeated atleast three times and the average mass of sand filling the cone is found (M_2).

v. Internal dimensions and volume of the calibration can be measured. The can is filled with water up to the brim and the mass is found. From the mass the volume is found and checked with the previous one and average volume, V, is determined.

vi. The pouring cylinder is placed concentrically on top of the calibration can with initial mass, M_1 The shutter is opened and the sand is allowed to fill the can. The cylinder is tapped so as to ensure that the can and the conical portion are completely filled with sand. The cylinder is weighed.

vii. The step vi is repeated thrice and the average mass M_3 of the cylinder after filling the cone and the can is recorded.

viii. An area of about 450 mm² is cleaned and levelled.

ix. The square tray with a central hole is placed on the prepared surface. A circular hole of 100 mm diameter and 150 mm depth is roughly dug. All the soil are carefully collected and the mass is found as (M_s). In case of a fine-grained soil a core-cutter is pushed approximately up to a depth of 100 mm, the soil is collected and weighed as M_s. The core-cutter is kept in position till the rest of the procedure.

x. Some representative soil for water content determination are taken.

xi. The tray is removed and the pouring cylinder is concentrically placed on the hole with initial mass M_1. The shutter is opened and the sand is allowed to fill the core and hole. The cylinder is weighed as (M_4).

3. Computation and Result

Mass of sand filling calibration can = M_a

$$= (M_1 - M_3 - M_2)\ g$$

Bulk density of sand, $\rho_{sd} = \frac{M_s}{V}$ g/cc

Mass of sand required to fill the excavated hole, $M_b = (M_1 - M_4 - M_2)$ g

Bulk density of soil, $\rho = (M_s / M_b) \times \rho_{sd}$ g/cc

Dry density of soil, $\rho_d = \frac{\rho}{1+\frac{w}{100}}$ g/cc

The dry density and water content of the soil are reported to the second decimal place and to two significant figures.

4. Applications and Limitations

In granular soils with a little or no fines there is a possibility of slumping of the sides of the excavated soil in the hole. In order to avoid bulking of sand due to atmospheric humidity, it is necessary to check the bulk density during each day's work.

For fine and medium-grained soils, a three-litre pouring cylinder is used whereas for fine, medium and coarse-grained soils a greater capacity of 16.5 litres is used. As the in-place density of a soil varies from point to point, tests are conducted at different locations close to each other and the mean value is recommended.

4.2 California Bearing Ratio

California Bearing Ratio Test referred as CBR Test is an adhoc penetration test developed by the California State Highway Department of the USA. It is used for the evaluation of subgrade strengths for roads and pavements.

It is expressed as the percentage of force per unit area required to penetrate a soil mass with a circular plunger of 50 mm diameter at a rate of 1.25 mm/min compared with that required for the corresponding penetration in a standard material.

Laboratory Method

1. Apparatus Required

i. A mould shall be made of non-corrodable metal and be cylindrical in shape with an inside diameter of 150 mm and of height 175 mm. It shall be provided with a detachable metal extension collar of 50 mm in height and a detachable perforated base plate of 10 mm thick. The perforations in the base plate shall not exceed 1.5 mm in diameter.

ii. A collar of 50 mm height and 150 mm diameter.

iii. Circular metal spacer disc of 148 mm diameter and 47.7 mm in height.

iv. Metal rammer having a weight of 2.6 kg with a drop of 310 mm or having a weight of 4.89 kg with a drop of 450 mm.

v. Expansion measuring apparatus consists of a plate 148 mm in diameter with perforations of 1.5 mm in diameter with a threaded stem at the center. A metal tripod fitting over the rim of the mould with a dial gauge clamped for measuring the amount of swell during soaking.

vi. Annular metal weight and several slotted weights weighing 2.5 kg each.

vii. Loading machine with a capacity of 50 kN (5000 kg approximately) with a rate of strain of 1.25 mm/min.

2. Preparation of Test Specimens

The test may be performed on:

i. Undisturbed specimens or

ii. Remoulded specimens which may be compacted either statically or dynamically.

Undisturbed Specimen

This is obtained by fitting a cutting edge of 150 mm diameter to the mould and pushing the mould as gently as possible into the ground. As the mould is pushed in, the soil is dug from the outside. When the mould is full of soil, it is removed by under-digging.The top and bottom surfaces are trimmed flat so as to get a specimen of the required length ready for testing. If the soil is hard and the mould cannot be pressed, then a large undisturbed lump of soil is cut out, from which the required specimen for the mould is made. If the specimen is loose in the mould, the annular cavity is filled with paraffin wax. The density of the soil and the water content are determined so as to determine the dry density.

Remoulded Specimen

The dry density of the remoulded specimen has to be decided which may be the field density, maximum dry density or any other density. Accordingly the moulding water content may be OMC or field moisture content. Soil for the remoulded specimen shall pass through a 19 mm sieve.

Allowance for large size particles is made by replacing the soil by an equal amount of material passing a 19 mm sieve and retaining in a 4.75 mm sieve. Soil is prepared and compacted in the mould either statically or dynamically as required.

In both cases of compaction if it is intended that the specimen is to be soaked, the water content of the soil before and after compaction is determined. If the specimen is not to be soaked, a representative sample is taken for water content determination.

3. Test Procedure

i. Test for Swelling

A filter paper is placed over the specimen and the adjustable stem and the perforated plates are placed on the soil specimen. Weights are then applied to produce a surcharge equal to the weight of the base material and pavement expected in actual construction within 2.5 kg. Whole weight should not be less than 5 kg.

Whole mould and weights are immersed in a tank of water allowing free access of water to the top and bottom of the specimen. Tripod for the expansion measuring device shall be mounted on the edge of the mould and the initial dial gauge reading recorded.

Set-up is kept as such without any disturbance for 96 hours. Readings are noted down every day against the time of reading. During this entire period water level in the tank is maintained.

At the end of the soaking period, change in the dial reading is noted. The tripod and the mould are removed from the water tank. The specimen is allowed to drain for 15 minutes downwards. Weights, perforated plate and filter paper are removed. Weight of the mould along with the soaked soil specimen is found.

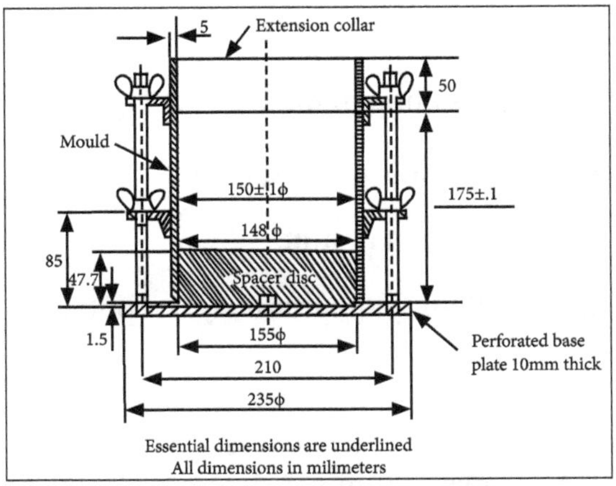

Typical mould with extension collar.

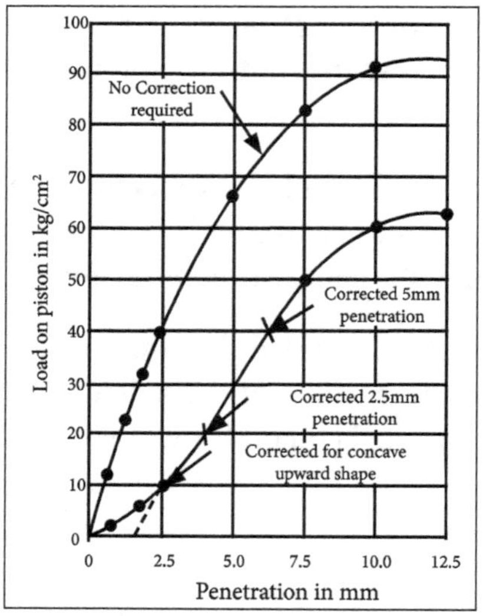

Correction of load penetration curves.

ii. Penetration Test

The mould along with the specimen is placed on the testing machine. Required number of surcharge weights to simulate the intensity of loading equivalent to the base material and pavement is placed. Plunger is set to touch the specimen with a seating load of 4 kg.

Load on the plunger is applied at the rate of 1.25 mm/min. Loads are noted at different penetration levels till 12.5 mm. Soil samples are taken for moisture content determination. The test may be conducted on the rear side of the specimen.

4. Computation and Result

i. Expansion Ratio

$$\text{Expansion ratio} = \frac{d_f - d_s}{h} \times 100$$

where,

d_f is the final dial gauge reading in mm

d_s is the initial dial gauge reading in mm and

h is the height of the specimen in mm.

The expansion ratio is used to identify qualitatively the potential expansiveness of soil.

ii. CBR Value

The load penetration curve is plotted. The curve is usually convex upwards although the initial portion of the curve may be convex downwards due to surface irregulations.

A correction shall be made. The CBR values are usually calculated for penetrations of 2.5 and 5 mm.

Then, $CBR = \frac{P_T}{P_s} \times 100$

where,

P_T is the corrected unit (or total) test load corresponding to the chosen penetration from the load-penetration curve and

P_s the unit (or total) standard load for the soil depth of penetration as for P_T taken from below table.

Table: Standard loads.

Penetration depth, mm	Unit standard load, kg/cm²	Total standard load, kg
2.5	70	1.370
5.0	105	2.055
7.5	134	2.630
10.0	162	3.180
12.5	183	3.600

Generally, the CBR value at 2.5 mm penetration will be greater than that at 5.0 mm penetration. In such a case, the former shall be taken as the CBR value for design purposes. If the CBR corresponding to a penetration of 5 mm exceeds that of 2.5 mm, the test shall be repeated.

4.2.1 Modulus of Subgrade Reaction

Modulus of Subgrade Reaction (K_s) is measured and expressed as load intensity per unit of displacement. It is often expressed in $kip/in^2/in$. In the SI system, it is expressed as $kN/m^2/m$. Some express this term in kip/in^3 which may be misleading. Numerically, kip/in^3 represents the physical significance of the measured value and could be mistaken as a density unit or a volumetric measurement.

Mathematically, the coefficient of subgrade reaction is expressed as:

$$K_s = p/s$$

where,

p = contact pressure intensity

s = soil settlement

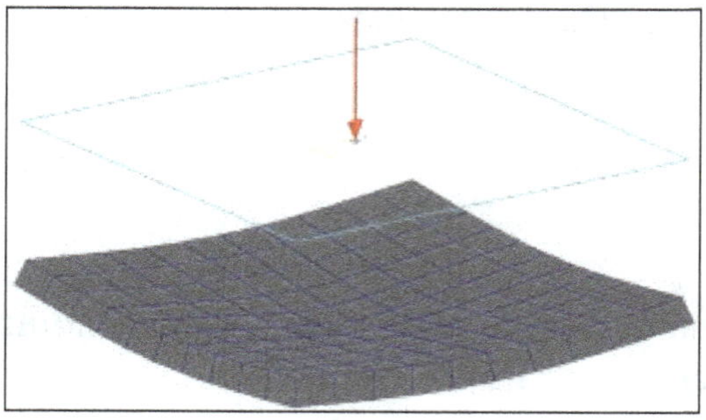

Soil pressure contour.

The term subgrade reaction refers to the pressure distribution which is the result or 'reaction' of the 'subgrade' to a load imposed upon top of a foundation structure. In practice, the foundation structure is commonly a reinforced concrete slab or mat, and the subgrade usually refers to the soil or rock upon which the structure is constructed.

The selection of coefficient of subgrade reaction "k" for the design of rafts, beams, and rigid pavements is not satisfactory because of wide range of values recommended in the literature for each soil type and density. A simple relationship was found between k and the standard penetration test values that takes into account the effect of the footing size or width of the loaded area. Test results are compared with other plate load tests on granular soils in different areas around the world, to establish confidence in the derived correlation and to explain the causes of variation in different granular deposits.

Determination of the Modulus of Subgrade Reaction

The modulus of subgrade reaction and its probabilistic characteristics can be determined directly by field test. As an alternative to the determination of the modulus of subgrade reaction, regular soil properties can be used.

In 1955, Karl Terzaghi (Liao, 1995) published a classic paper titled 'Evaluation of coefficients of subgrade reaction', in which he presented recommendations for estimating the spring constants that have come to be commonly used to model the foundation subgrade in the analysis of mat foundations and other similar problems.

One of the popular models in determining the modulus of subgrade reaction is Winkler model. In this model the subgrade soil assumes to behave like infinite number of linear elastic springs that the stiffness of spring is named as the modulus of subgrade reaction. This modulus is dependant to some parameters like soil type, size, shape, depth and type of foundation, etc.

The direct method to estimate the modulus of subgrade reaction is plate load test that it is done with 30-100 cm diameter circular plate or equivalent rectangular plate (Reza & Janbaz, 2008). Because of limitation of available data and the uncertainty of soil condition, it was also proposed to use the empirical equations. The empirical equations given below (Widjaja, 2008) were suggested to estimate k value both in clay and sand:

Clay: $k = 40 - 50\ s_u\ (t/m^3)$

Sand: $k = 70 - 100\ NSPT\ (t/m^3)$

where s_u is an undrained shear strength (t/m^2) and NSPT is a value of Standard Penetration Test.

Effect of the Modulus of Subgrade Reaction

In the present work, the effect of soil modulus is investigated by fixing all the other parameters and varying the modulus of subgrade reaction from 60 MN/m^3 to 120 MN/m^3 for same typical and symmetric raft foundation.

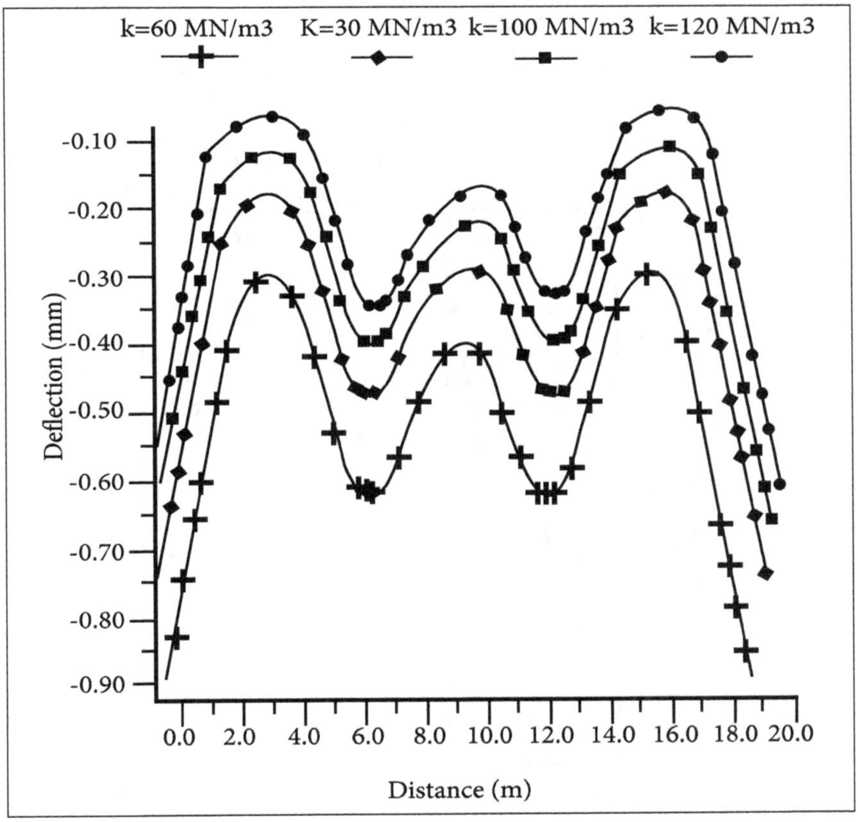

Deflection vs. distance with k variable.

When studying the effects of soil modulus the following values are used:

L = 6.0 m, d = 0.5 m, column dimensions: 0.4 m x 0.4 m, column load = 1.0 MN, modulus of elasticity for the raft material = 21 GPa, thickness (t) = 0.6 m and the modulus of subgrade reaction for supporting soil (k) is variable.

Figure represents the deflected shape of the raft for a section directly under an interior columns line, while the figures below show the bending moment and shear force diagrams respectively, for an interior columns strip of the raft, with various values of (k).

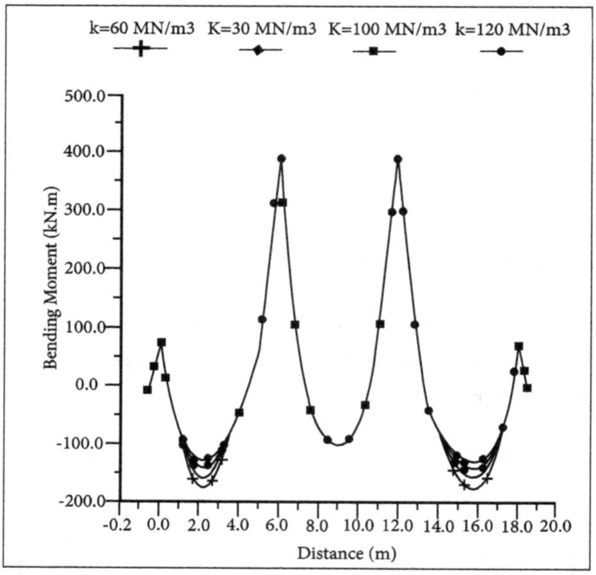

Shear force vs. distance with k variable.

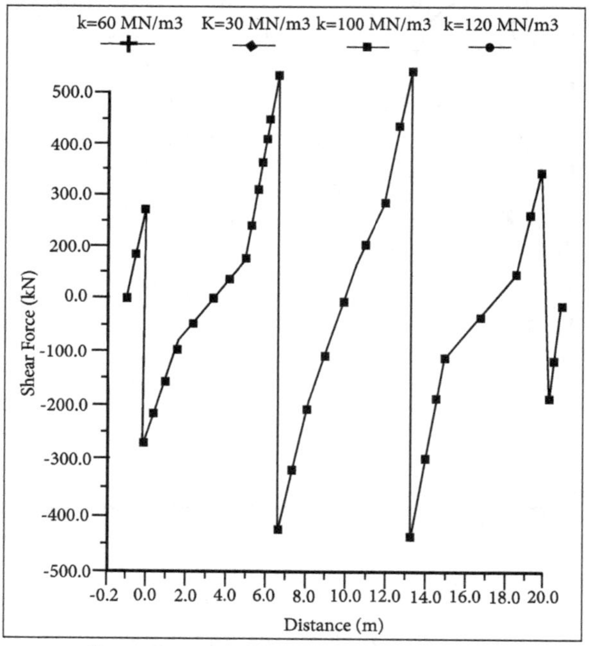

Shear force vs. distance with k variable.

4.3 Stone Aggregates

Aggregates form the major portion of the pavement structure. Bear stresses occurring on the roads and have to resist wear due to abrasive action of traffic. Aggregates are also used in the flexible as well as in rigid pavements. Therefore, the properties of aggregates are of considerable importance to the highway.

Desirable Properties of Road Aggregates

1. Hardness
2. Strength
3. Durability
4. Toughness
5. Adhesion with bitumen
6. Shape of aggregates

Hardness

The aggregates used in the surface course are subjected to constant rubbing or abrasion due to moving traffic. Abrasive action may be increased due to the presence of abrasing material like sand between tyres of vehicle and the aggregates exposed to the top surface. Thus, they should be hard enough to resist the wear due to abrasive action of traffic.

Strength

The aggregates to be used in road construction, particularly the aggregates used in wearing course of the pavement should be sufficiently strong/resistant to crushing to withstand the high stresses induced due to heavy traffic wheel loads.

Durability

The aggregates are subjected to physical and chemical actions of the rains and ground water, the impurities in them and that of atmosphere. Thus it is desirable that the road stones used in the construction should be sound enough to withstand the weathering action. The property of the aggregates to withstand the adverse actions of weather may be called soundness.

Toughness

Aggregates in the pavement are also subjected to impact due to moving wheel loads. The magnitude of impact increase with roughness of road and the speed of vehicle. Severe impact is common when heavily loaded steel tyred vehicles move on WBM. The resistance to impact or toughness is thus another desirable property of aggregates.

Adhesion with Bitumen

The aggregates in bituminous pavements must have less affinity with water when compared with bitumen, otherwise the bituminous coating on the aggregates will be stripped off in presence of water.

Shape of Aggregates

Aggregates may be rounded, angular, flaky or elongated. Flaky and elongated particles have less strength than rounded and cubical particles. Thus, too flaky and too much elongated particles should be avoided.

4.3.1 Tests for Road Aggregates

Tests for road aggregates to check the desirable properties are to be conducted. The relevant tests are:

i. Crushing Test

ii. Abrasion Test:

  a. Los Angeles Test

  b. Deval's Test

iii. Impact Test

iv. Soundness Test

Crushing Test

1. Theory

The aggregate crushing value gives a relative measure of the resistance of an aggregate to crushing under a gradually applied compressive load. The standard aggregate crushing test is made on aggregates passing a 12.5 mm IS sieve and retained on 10 mm IS sieve.

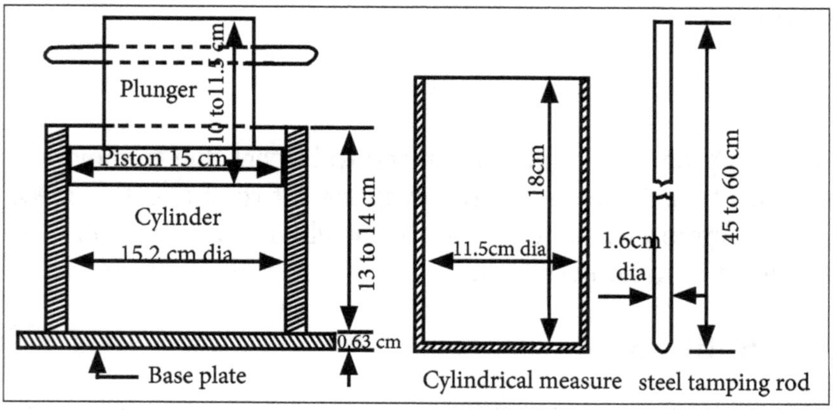

Aggregate crushing test apparatus.

2. Apparatus Required

i. A steel cylinder of 15.2 cm diameter and 13 to 14 cm height with a base plate and a plunger

ii. Cylindrical measure of 11.5 cm diameter and 18 cm height

iii. Tamping rod of 1.4 cm diameter and 45 to 50 cm long

iv. IS standard sieves

v. Compression testing machine

3. Test Procedure

i. Dry aggregate of 6.5 kg passing 12.5 mm IS sieve and retained on 10 mm IS sieve is taken.

ii. The aggregate is filled into the standard cylindrical measure in three layers approximately of equal depth.

iii. Each layer is tamped 25 times with the tamping rod and finally levelled off using the tamping rod as straight edge.

iv. The weight of the sample contained in the cylindrical measure is measured (M_1). The same weight of the sample is taken for the subsequent test.

v. Then the test sample is placed in the test cylinder in three equal layers tamping each layer 25 times.

vi. Then the plunger is placed on the top of the sample along with the cylinder which is in turn placed on the compression testing machine.

vii. The sample is loaded uniformly at a rate of 4 tonnes per minute up to a load of 40 tonnes.

viii. The load is then released and the whole of the material is removed from the cylinder and sieved on 2.36 mm IS sieve.

ix. The material passed through the sieve is weighed (M_2).

4. Computation and Result

The aggregate crushing value is the percentage of the crushed material passing 2.36 mm IS sieve in terms of the original weight of the specimen.

$$\text{Aggregate crushing value} = \frac{M_2}{M_1} \times 100$$

5. Applications and Limitations

i. Aggregates with low aggregate crushing value is the strong one.

ii. For a base course the recommended aggregate crushing value should not exceed 45% and for wearing surface courses should be within 30%.

iii. For concrete the aggregate crushing value should not be more than 45%.

iv. If required, or if the standard size is not available, other sizes up to 25 mm may be tested. Because of non-homogeneity of aggregates the results will not be comparable with those obtained in the standard test.

Aggregate Abrasion Tests Theory

Aggregates used in the surface course are subjected to wear and tear which is totally referred as abrasion. Abrasion value of aggregates is the difference between the original weight and the final weight of the test specimen tested in a standard abrasion testing machine and represented as a percentage. Indian Standard 2386 (Part IV) of 1963 recommended the use of Los Angeles abrasion testing machine or the Deval's abrasion testing machine.

Los-Angeles Abrasion Test

1. Equipment Required

Los-Angeles abrasion test is based on the principle that the percentage wear is due to the relative rubbing action between the aggregates and steel balls used as abrasive charge. The weight of each charge and the number required depends on the type of grading of test samples.

Table shows the grading of test samples A,B,C,D,E,F and G and Corresponding specified abrasive charge to be used is given in table. Each charge consists of cast iron spheres or steel spheres approximately 48 mm in diameter and each weighing between 390 to 445 gm.

Table: Gradings of test samples.

Sieve size		Weight in gm of test sample for the grade						
Passing mm	Retained mm	A	B	C	D	E	F	G
80	63	-	-	-	-	2500	-	-
63	50	-	-	-	-	2500	-	-
50	40	-	-	-	-	5000	5000	-
40	25	1250	-	-	-	-	5000	5000
25	20	1250	-	-	-	-	-	5000

20	12.5	1250	2500	-	-	-	-	-
12.5	10	1250	2500	-	-	-	-	-
10	6.3	-	-	2500	-	-	-	-
6.3	4.75	-	-	2500	-	-	-	-
4.75	2.36	-	-	-	5000	-	-	-

Table: Specified abrasive charge.

Grading	Number of spheres	Weight of charge (gm)
A	12	5000 ± 25
B	11	4584 ± 25
C	8	3330 ± 20
D	6	2500 ± 15
E	12	5000 ± 25
F	12	5000 ± 25
G	12	5000 ± 25

Los Angeles machine-It consists of a hollow cylinder closed at both ends, having inside diameter 70 cm and length 50 cm and mounted so as to rotate about its horizontal axis.

2. Test Procedure

i. The specified weight of aggregate sample (5 or 10 kg) depending on the gradation is placed in the machine.

ii. The abrasive charge corresponding to the specified weight of aggregate is added to the machine.

iii. The machine is rotated at a speed of 20 to 33 rev/min.

iv. For grading A, B, C and D, the machine is rotated for 500 revolutions.

v. For grading E, F and G it is rotated for 1000 revolutions.

vi. After the completion of the above number of revolutions the material is discharged from the machine.

vii. Preliminary separation of the sample is made on a sieve coarser then 1.7 mm sieve. Finer portion is than sieved on a 1.7 mm IS sieve.

viii. The material coarser than 1.7 mm IS sieve is washed, dried in an oven at 105° to 110°C to a substantially constant weight and accurately weighed to the nearest gram.

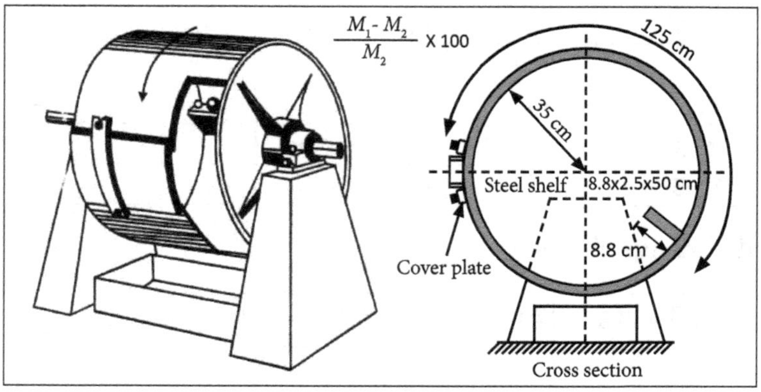

Los Angeles abrasion testing machine.

3. Computation and Result

The difference between the original weight (M_1) and the final weight of the test sample (M_2) is expressed as a percentage weight of the test sample.

$$\text{Los} - \text{Angeles Abrasion Value} = \frac{M_1 - M_2}{M_2} \times 100$$

4. Applications and Limitations

i. The percentage of wear should not be more than 16% for concrete aggregate.

ii. For bituminous mixes Los-Angeles abrasion values up to 30% is allowed.

iii. For base courses like bitumen bound macadam Los Angeles abrasion value up to 50 % is allowed.

iv. This test is more dependable as rubbing and pounding action taking place inside the machine almost simulates the field condition.

v. Further correlation of Los Angeles abrasion value with field performance are already established.

Deval Abrasion Test

1. Equipment Required

i. It's principle is almost same as that of Los Angeles abrasion test.

ii. Here the samples of aggregates are allowed to tumble over in a rattler in the presence of abrasive charge. The abrasive value is expressed as in the Los Angeles abrasion test.

iii. Here the abrasive charges consist of 6 cast iron or steel sphere of about 4.8 cm diameter and total weight is 2500 g.

iv. Deval machine consists of two hollow cylinders of 20 cm diameter and 34 cm length mounted in such a way that the cylinders rotate about the horizontal axis. The axis of the cylinders is 30° angle with the horizontal.

2. Test Procedure

i. Specified quantity of aggregate samples (4 to 5.5 kg) of any one of the specified gradings is placed in the cylinder.

ii. Two tests are conducted simultaneously as two cylinders are available.

iii. The machine is rotated at a speed of 30 to 33 rpm for 10,000 revolutions.

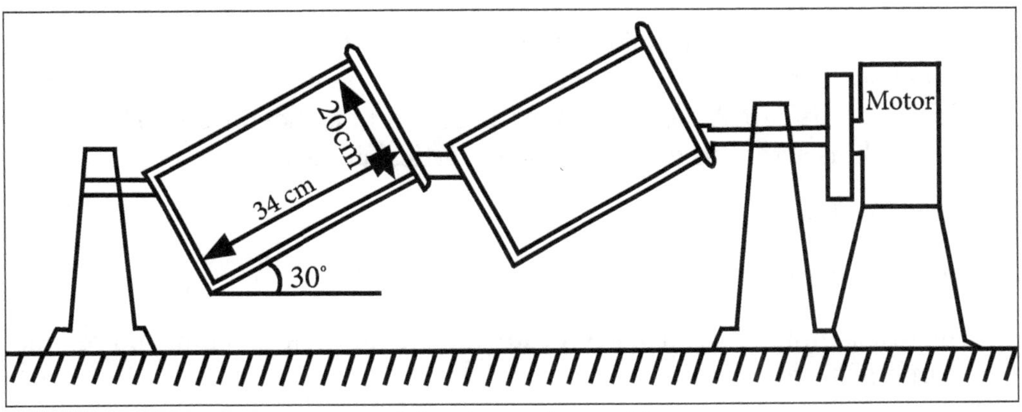

Deval abrasion testing machine.

iv. After completion of revolutions the material is sieved through 1.7 mm IS sieve and washed, dried and weighed.

3. Computation and Result

$$\text{Deval Abrasion Value} = \frac{M_1 - M_2}{M_1} \times 100$$

where,

M_1 is the initial weight of the sample

M_2 is the final weight of the sample

As two samples are tested at the same time better average value can be obtained. When the test is carried out without charges the test is known as Deval attrition test.

Aggregate Impact Test

1. Theory

It is conducted to evaluate the toughness of stones or resistance of stones to breaking under repeated impact loads. It may be carried out either as a cylindrical stone specimen or stone aggregates. The Page Impact Test is one which has been designed to evaluate the toughness of stone specimens.

The impact test conducted on aggregate is known as Aggregate Impact Test which is carried out to evaluate the resistance to impact of aggregates.

The aggregate impact value is different from the crushing test in which the test value indicates a relative measure of resistance of aggregate to shock or impact rather than a gradually applied compressive load.

2. Equipment Required

i. Aggregate Impact Testing Machine: It consists of a metal base and a cylindrical cup of 10.2 cm internal diameter and 5 cm in depth. The aggregate sample is placed in this cup. A metal hammer of 13.5 to 14 kg having a free fall from a 38 cm height is arranged to drop through vertical guides.

ii. Cylindrical Measure

3. Test Procedure

i. The test sample consists of aggregate passing through 12.5 mm and retained on 10 mm IS sieve.

ii. The aggregate is dried in an oven for a period of four hours at a temperature of 100° to 110° C.

iii. The sample is filled in the cylindrical measure in 3 layers by tamping each layer by 25 blows.

iv. The net weight of the aggregate in the cylindrical measure is determined (M_1). This weight of aggregate has to be used for duplicate test on the same material.

v. The whole sample is transferred to the cup of the aggregate impact machine and compacted by tamping 25 times.

vi. The hammer is raised to a height of 380 mm above the upper surface of the cup and allowed to fall freely on the aggregate sample in the cup.

vii. The test sample shall be subjected to a total 15 such blows each being delivered at an interval of not less than one second.

viii. The crushed aggregate is recovered from the cup and the whole of it is sieved on 2.36 mm IS sieve. The fraction passing the sieve is weighed to an accuracy of 0.1 gm (M_2).

ix. The fraction retained on the sieve is also weighed (M_3).

x. If the total weight ($M_2 + M_3$) is less than the initial weight M_1 by more than one gm the result shall be discarded and a fresh test is made. Two tests are made.

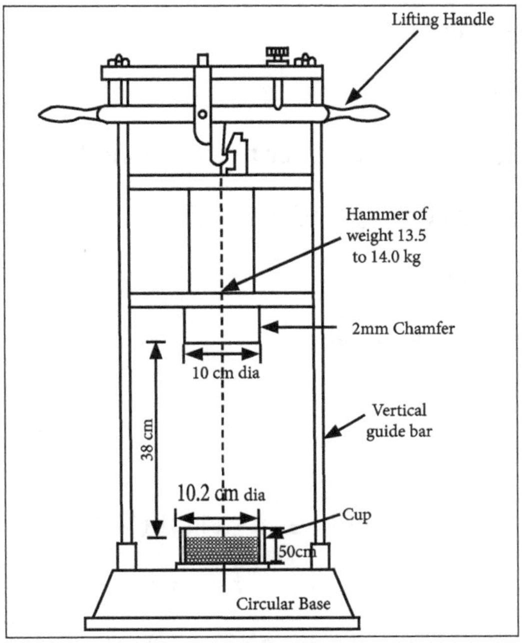

Aggregate impact testing machine.

4. Computation and Result

The ratio of the weight of fines formed to the total sample weight in each test is expressed as percentage. Therefore,

$$\text{Aggregate Impact Value} = \frac{M_2}{M_1} \times 100$$

5. Applications and Limitations

i. The aggregate impact value should not be more than 45 % for aggregates used for concrete.

ii. Concrete wearing surfaces should use aggregate impact value around 30%.

iii. Classification of stones are made on the basis of toughness property considering the aggregate impact values.

iv. Aggregate impact values less than 10 % is considered as exceptionally strong 10 to 20 % strong and 20 to 30 % satisfactory for road surfacing.

Aggregate Soundness Test

1. Theory

Soundness refers to the ability of aggregate to resist excessive changes in volume as a result of change in physical conditions. The physical conditions that affect the soundness are as follows:

i. Freezing and thawing

ii. Variation in temperature

iii. Alternate wetting and drying under normal conditions and

iv. Alternate wetting and drying under salt water.

2. Apparatus Required

i. Sodium or magnesium sulphate

ii. Weighing machine

3. Test Procedure

i. Carefully graded aggregate of known weight (M_1) is taken.

ii. The sample is soaked in a solution of magnesium sulphate for 16 to 18 hours.

iii. The sample is dried in an oven at 105 to 110° C to a constant weight. This completes one cycle.

iv. After the final cycle the sample is examined visually for the evidence of excessive splitting, crumpling or disintegration of grains. Steps iii and iv are repeated.

v. Sieve analysis is carried out to note the variation in gradation from the original.

4. Computation and Result

The average loss in weight is found which represents the soundness of the aggregate.

The average loss in weight should not exceed 12 or 18% when tested with sodium sulphate or magnesium sulphate respectively.

4.4 Bituminous Materials

Bituminous roads are constructed with varying aggregate sizes and compositions. This type of construction is adopted as a wearing course or a surface course. In heavy traffic roads bituminous construction are also adopted for base course.

Compared to cement concrete surfacing, bituminous surfacing takes very less time to use after construction. Black top construction is in large use in developing countries like India where cement is in great demand as a construction material. Further stage construction can be done effectively.

Types of Bituminous Construction

There are various types and methods of construction of bituminous pavements. The following construction techniques are in use:

i. Surface dressing or painting

ii. Interface treatments

iii. Premix materials like,

   a. Bituminous bound macadam

   a. Carpet

   b. Bituminous concrete

   c. Sheet asphalt or rolled asphalt

   d. Mastic asphalt

Bituminous binders adopted in the above methods are straight run bitumen, road tar, cutback or emulsion. The choice of type of binder is based on the type of construction, availability of material, climatic conditions, equipment, etc.

For example, bitumen and tar require heating to bring them to a proper viscosity for their use. Use of this type is called hotmix technique. On the other hand cutback and emulsion do not require heating as that of bitumen or tar. Use of this material is called cold-mix technique. Based on the technique of mixing and construction, the bituminous constructions are classified as,

1. Road-mix methods.
2. Central-plant mix methods.

Premix materials construction depends on the gradation. They are classified as open-graded and dense-graded.

Bitumen construction method for base course are grouted or penetration macadam (either fully-grouted or semi-grouted) and bitumen bound macadam. Different bituminous constructions employed in surface course of pavements are surface dressing, grouted or penetration macadam and seal coat and bituminous premix surfacing.

Surface Dressing

1. Functions and Applications

Main functions of the surface dressing are:

i. To provide a dust free pavement surface over the base course.

ii. To provide full stability to the base course and pavement structure as a whole.

iii. To prevent infiltration of surface water and thereby form as a water proof layer.

In water-bound macadam roads the binding material, soil slurry, is removed easily by fast moving vehicles. They expose the stone pieces and also make them loose. In such situations a surface dressing by bitumen for a thickness of 2 to 3 cm improves the surface performance better and adds to the overall stability of the structure.

2. Materials Requirement

The bitumen binder may be bitumen or tar or cutback. For a WBM road or renewal of black top surface, the weight required for 10 sq. m of area are given in table.

Table: Binder requirement.

Base course type	Binder requirement in kg per 10 sq. m. area					
	Single or first coat			Second coat		
	Bitumen	Tar	Cutback	Bitumen	Tar	Cutback
WBM	17 to 19.5	17 to 22	19 to 22	10 to 12	12 to 15	12 to 15
Renewal of black to surface	10 to 12	12 to 17	10 to 12	10 to 12	12 to 15	12 to 15

3. Construction Procedure

Surface dressing can be done on an existing road surface as a black top surface or as a new WBM road. In the case of new WBM road surface, the metals are properly rolled after sprinkling water and then the dressing is applied. In case of an old road or existing WBM, the surface is reconditioned to have the design camber and defects rectified.

The surface is cleaned with wire brush such that the aggregates are exposed atleast 1.25 cm. Surface dressing is done in one or two coats depending on the necessity. Over the prepared surface a uniform layer of bituminous binder is applied as given in table.

In general 2 kg of binder per square meter surface is sufficient. In order to avoid bleeding no excess binder should be used. After spraying, binder chippings of 12 mm thick (for first coat) of 12 mm size of 0.14 to 0.15 cm of materials for 10 sq. m area is used. For second coat 9 mm thick of 9 mm size of 0.09 to 0.11 cm of material for 10 sq. m area is applied.

Rolling is started after this from the edges towards the center longitudinally with an overlapping not less than half of the width of rolling wheel. After completion of half of the surface the rolling is started from the other edge. The rolling is continued till all the particles are properly interlocked. This forms the final rolling in case of single coat.

In the case of additional coat, the binder is applied to the prepared surface and chippings spread as per requirement and rolled finally. After ensuring proper cross-sections and 24 hours after rolling the road may be opened to traffic.

Interface Treatment

1. Functions and Applications

Depending on the condition of the existing base the following interface treatments are recommended:

1. Prime coat is the first coat of bituminous surfacing done on an existing previous texture base.
2. Tack coat is the bituminous surfacing done over an existing cement concrete top or already existing black top surface road.
3. Seal coat is a final thin coat over certain previous bitumen pavements. The functions of a seal coat are:
   i. To increase the skin resistance of the surface
   ii. To make the surface water proof and
   iii. To increase the life of the surface.

Penetration of grouted macadam is the process by which the bitumen is spread over the compacted aggregate surface to allow the bitumen to fill voids and binding the aggregates together. When the bitumen penetrates fully it is called full grout.

Generally, the above treatments are given to existing WBM road surfaces. It is essential that adequate bond should exist between the newly laid bituminous layer and the existing surface at the interface.

2. Construction Procedure

The existing surface is lightly brushed and reconditioned. Priming coat is applied if required.

Additional course aggregates are spread with proper protection of edges. The profile of the cross section is checked with template. A 10 ton roller is used to roll the aggregates when dry.

Rolling is continued till crushing of aggregates are ensured. The profile of the dry-rolled surface is checked for its correctness. Over the dry and compacted surface, the specific binder (bitumen) is uniformly spread either with pressure distributor or hand sprayer.

For a compacted surface of one sq. m with a thickness of 5 cm and 7.5 cm, 50 kg and 68 kg bitumen respectively are required. After this process, chippings or key aggregates are spread and rolled. Key aggregates of 10 to 12 mm size is used. Volume of material needed varies from 0.15 to 0.18 m^3 respectively.

Over this surface a premixed sand bitumen or surface dress type seal coat is applied after a few days or immediately.

4.4.1 Tests on Bitumen

Tests on Bituminous Materials

Bitumen is available in a variety of types and grades. To assess the desirability condition different tests are discussed in the subsequent sections. Different tests which are generally used are:

1. Penetration test
2. Ductility test
3. Viscosity test
4. Softening point test

For classification of bitumen and for the study of performance of bituminous pavements, penetration and ductility tests are needed. Softening point, flash and fire point tests are essential to guide the paving technologists on the field. Each test has the limitations as bitumen from different sources behave differently.

For example, bitumen from different sources may have same penetration at a particular temperature but may exhibit different viscous characteristics. For a particular test, value can be true for the given sample only which cannot be correlated with any other fundamental property.

Penetration Test

1. Theory

Hardness or softness of bitumen is assessed by penetration of a standard needle.

It is obtained by measuring the depth in tenth of a millimetre to which a standard loaded needle shall penetrate vertically in 5 seconds.

The bitumen grade is specified in terms of penetration value.

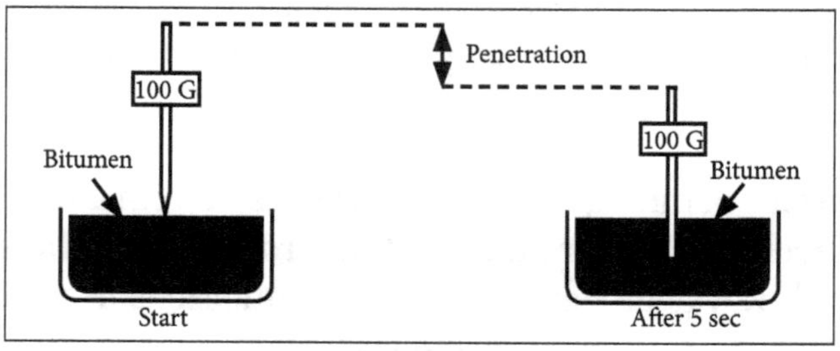

Penetration test concept.

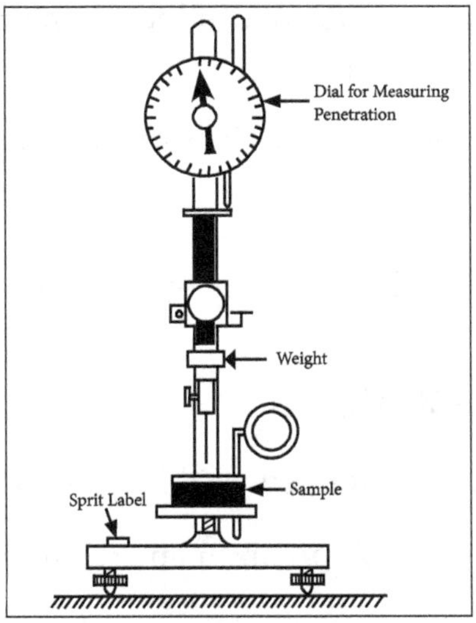

Penetrometer

2. Apparatus Required

i. Penetrometer consists of a needle with a device to release and lock a 100 g weight.

   A graduated dial is provided to read the penetration values up to an accuracy of $1/10^{th}$ of a millimetre.

ii. A sample container

iii. Temperature controlled water bath maintained at 25° C

iv. Benzene liquid

3. Test Procedure

i. The bitumen is adequately softened such that it attains pouring consistency.

ii. The softened material is stirred thoroughly and poured into the sample containers to a depth atleast 15 mm in excess of the expected penetration.

iii. The containers are placed in the water bath for one hour.

iv. The sample container is taken out of the bath and kept on the machine and the needle is made to contact with the surface of the sample.

v. The dial is set to zero or the initial reading is taken and the needle is released for 5 seconds. The final reading is taken. The depth of penetration is reported in one-tenth millimetre units.

vi. After each test the needle is disengaged and the needle point is wiped with benzene and dried.

vii. Atleast three penetration tests are conducted on the same sample by testing at distances of atleast 10 mm apart.

viii. The average value of these three tests is reported as penetration value.

4. Computation and Result

The bitumen grade is reported with the range of penetration value, e.g., 80 - 100 or 80 /100.

5. Applications and Limitations

i. The penetration values of bitumen used in general varies from 20 to 225.

ii. The grades 30 / 40 and 80 / 100 bitumen are more commonly used.

iii. In hot weather condition a penetration grade bitumen like 30 / 40 is preferred.

iv. The penetration test is conducted only for bitumen.

v. The penetration value is very much affected by pouring temperature, size of needle, weight placed on the needle and test temperature.

Ductility Test

1. Theory

Ductility is the property of bitumen which permits the material to undergo great deformation without breaking. It is expressed as the distance in centimeters to which a standard briquette of bitumen of 1 sq cm can be stretched without breaking when the rate of pull is 5 cm per minute at the test temperature of 25°C to 27°C.

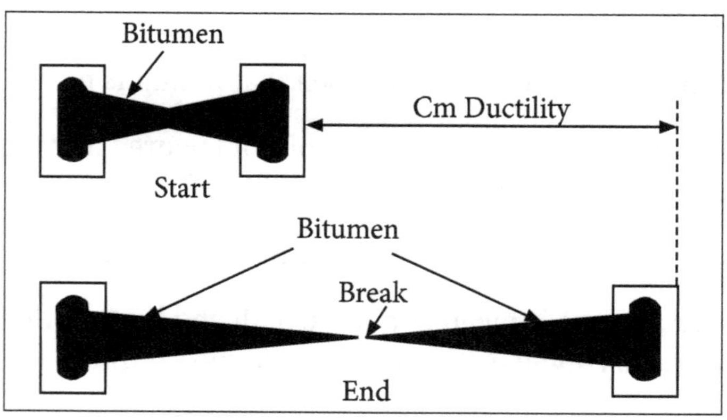

Ductility test

2. Apparatus Required

i. Ductility test apparatus consists of two briquette mould which can be moved at a pre-calibrated rate.

ii. Constant temperature water bath.

3. Test Procedure

i. The bitumen sample is made fluid by heating, poured into two-mould assembly and placed on a brass plate.

ii. At first, the sample and mould assembly is cooled in air and then in water bath maintained at 27°C.

iii. The excess bitumen material is removed and levelled using a hot knife.

iv. Then, the mould assembly and the sample is replaced in water bath for 85 to 95 minutes.

v. The sides of the mould are removed, the clips are hooked on to the machine. The pointer is adjusted to zero.

vi. One part of the assembly is allowed to pull at the rate of 5 cm per minute.

vii. The distance up to the point of breaking of thread is noted and the distance is reported in centimetres.

4. Computation and Result

The distance from start till the breaking of bitumen is reported in centimetres.

5. Applications and Limitations

i. Ductility test is conducted to find the binding property of a bitumen which serves as a satisfactory binder in flexible pavement.

ii. This test is a measure of the adhesiveness and elasticity of the bitumen.

iii. Bitumen should satisfy both the penetration and ductility for a good field work.

iv. For different bitumen, the ductility values vary from 5 to over 100.

v. For satisfactory performance of a bitumen, the value should not be less than 50.

vi. Ductility values get affected by factors such as pouring temperature, dimensions of briquette, level of briquette in water bath, pressure of air pockets, test temperature and rate of pulling.

Viscosity Test

1. Theory

It is a measure of resistance to flow. Higher the viscosity of a liquid signifies that it can be nearer to semi-solid state of consistency.

It is a function of temperature. It decreases with increase in temperature. Here the time taken by a sample, is measured for 50 ml of the material to flow through a standard orifice under standard falling head conditions at a standard temperature.

2. Apparatus Required

i. Orifice type viscometer

ii. Cup

iii. Timer

3. Test Procedure

i. The cutback bitumen in viscous form is placed in the viscometer

ii. The timer is started and flow of the liquid allowed which falls drop by drop in the standard 50 cc spherical glass bulb.

iii. The time taken in seconds to fill the specific mark in the bulb is noted and expressed as the viscosity.

4. Applications and Limitations

i. At low viscosity the bituminous binder simply lubricates the aggregate particles instead of forming an uniform film for binding action.

ii. High viscosity resists the compaction effort.

iii. Orifice type viscometer indirectly finds the viscosity.

iv. Furol viscosity is a specific test conducted in liquid bituminous materials.

v. Ten millimetre orifice viscometer is recommended for road tar.

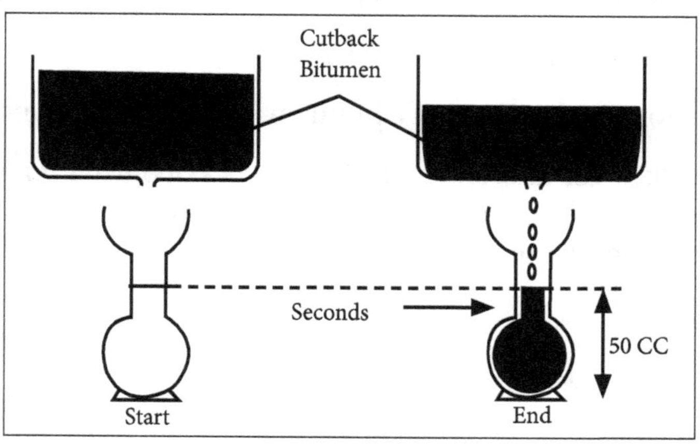

Viscosity test set-up.

vi. Viscosity is a function of temperature, it decreases with increase in temperature.

Softening Point Test

1. Theory

Softening point is the temperature at which the substance attains a particular degree of softening under specified condition of test.

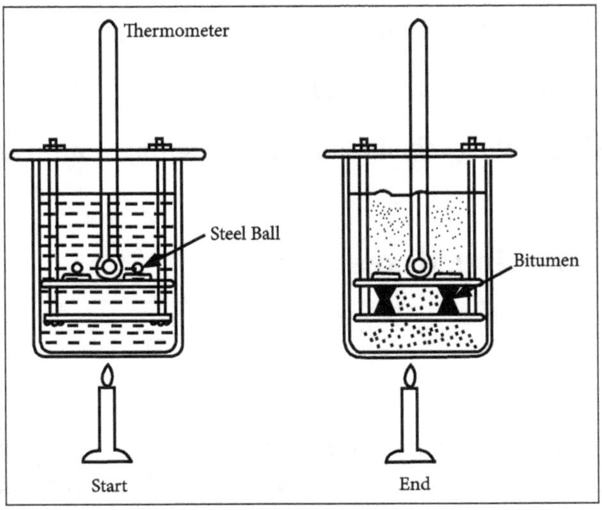

Softening point test set-up.

2. Apparatus Required

i. Ring and ball set up

ii. Glycerine

iii. Beaker

iv. Burner

v. Thermometer

3. Test Procedure

i. The given bitumen sample is taken in the brass ring.

ii. The brass ring along with the bitumen sample is suspended in water (or glycerine) at a particular temperature.

iii. A steel ball is placed upon the bitumen and the water (or glycerine) is then heated at a rate of 5°C per minute.

iv. The bitumen will start melting and start flowing down.

v. At a particular temperature,the softened bitumen touches the metal placed at a specified distance below the ring is recorded.

4. Computation and Result

The temperature noted is the softening point of the bitumen.

5. Applications and Limitations

i. Higher softening indicates lower temperature susceptibility and is preferred in areas where warm weather exists.

ii. Hard grade bitumen possesses higher softening point than soft grade bitumen.

iii. Bitumen used in paving job should have softening point varying between 35° to 70°C.

4.4.2 Bituminous Paving Mixes

Bituminous Carpet

1. Materials

Bituminous carpet is a premix prepared from 10 to 12 mm size chippings with sand and bitumen binder. Thickness varies from 2 to 2.5 cm. It is used for surface coarse layer. Sand-bitumen carpet mix containing no coarse aggregate is used as a wearing course.

2. Construction Procedure

The undulations and the pot-holes, if any on the existing road are reconditioned properly with pre-coated aggregates. In case of old bitumen or WBM road, a tack coat is applied on the surface. If the surface is made out of soft aggregate, a prime coat is applied.

Bitumen binder is heated to the specified temperature and tack or prime coat is applied. Chippings and bitumen are heated separately to the specified temperature and then mixed so as to form a homogeneous mix.

The mix is spread over the prepared surface with suitable rakes. The cross-section is once again checked before rolling. At a stretch, the premixed material is laid for a length of 15 meters and rolled.

Wheel surfaces are kept wet with the help of wet gunny bags or sprinkling water to avoid adhesion of mix on the wheels. Rolling is continued till it stopped showing wave in front of the roller. In low rainfall areas, a premixed medium coarse sand is applied on the carpet.

For a 10 sq. m area, 6.83 kg of binder and 0.06 m^3 of sand are required. In high rainfall areas, a liquid seal sprayed with a layer of chippings applied over the carpet. Chippings of 6 mm size of 0.091 m^3 per 10 sq. m area are used. Then the chippings are rolled with a light roller. The pavement may be allowed for traffic after 24 hours of the final rolling.

Bituminous Concrete

1. Functions and Materials

Bituminous concrete surfacing has the following functions:

i. It provides a resilient, dust-free, water proof and smooth road surface.

ii. It resists heavy and mixed vehicles in a better way.

iii. It has high durability and riding qualities as that of cement concrete pavements.

In this method coarse and fine aggregates along with bitumen binder is used to give a dense mix. Thickness depends on the traffic intensity and on the quality of base course. Aggregates and binders are separately heated and then mixed till all the aggregates are coated with bitumen. Mastic asphalt is also used as a binder.

Bituminous concrete surfacing is done in two layers:

i. Binder course

ii. Wearing course

The coarse to fine aggregate ratio is 2:1. Binder used is 7.5% of the mix by weight.

2. Construction Procedure

The existing base course is reconditioned atleast one week before laying the binder course. Then the bitumen course layer will be laid. This is also a hotmix process.

The hot mix is collected and spread over the prepared surface. The camber and thickness of bitumen layer is checked. The placed concrete is rolled by a roller at a speed not more than 5 km/hour. The number of passes required to attain the final desired thickness depends on the thickness.

The initial rolling is done using a 8 to 12 tonne roller followed by fixed wheel pneumatic roller of 15 to 30 tonnes. The wheels are kept damp with water or wet gunny bags.

Bitumen Bound Macadam

1. Functions

Bitumen Bound Macadam are used in the following areas:

i. In hilly regions, they resist frost action better than normal base courses.

ii. In regions of high water table, they act as impervious cut off.

iii. By using them, superior quality aggregates may be replaced by medium quality aggregates.

The binder and aggregates are mixed prior to placement and the finished thickness is kept from 5 to 7.5 cm. Size of aggregate varies with the thickness. For a 7.5 cm thick, the maximum size is limited to 37 mm.

2. Construction Procedure

The existing surface is reconditioned to proper cross-section and the surface is cleaned. On the prepared and cleaned surface, a thin layer of binder (prime or teak coat) is applied on a 10 sq. m surface and 4.0 to 7.5 kg of binder is to be used for black top surface or 7.5 to 10 kg for untreated WBM surfaces.

Aggregates of proper gradation and binder are separately heated at about 120°C and then mixed in a mixture. The mixture is placed on the already prepared surface and uniformly spread for the required thickness with rakes. The cross-section is again checked.

Rolling is done as early as possible after placing premixed material with a 8 to 10 tonne roller. The rolling is started from the edges and processed towards the center with uniform overlapping. Wheels of the roller should be kept wet while rolling so as to avoid sticking of mixed material on the wheels.

4.4.3 Marshall Method of Mix Design

The Marshall Mix Design method was originally developed by Bruce Marshall of the Mississippi Highway Department in 1939. The main idea of the Marshall Mix Design method involves the selection of the asphalt binder content with a suitable density which satisfies minimum stability and range of flow values.

The Marshall Mix Design method consists mainly of the following steps:

i. Determination of physical properties, size and gradation of aggregates.

ii. Selection of types of asphalt binder.

iii. Prepare initial samples, each with different asphalt binder content.

    For example, three samples are made each at 4.5, 5.0, 5.5, 6.0 and 6.5 percent asphalt by dry weight for a total of 15 samples. There must be at least two samples above and two below the estimated optimum asphalt content.

iv. Plot the graphs for the following variables:

    a. Asphalt binder content versus density

    b. Asphalt binder content versus Marshall stability

    c. Asphalt binder content versus flow

    d. Asphalt binder content versus air voids

e. Asphalt binder content versus voids in mineral aggregates

f. Asphalt binder content versus voids filled with asphalt

v. Find the asphalt binder content that corresponds to the air void content of 4 percent.

vi. Find the properties at this optimum asphalt binder content by reference with the graphs. Compare each of these values against design requirements and if all comply with design requirements, then the selected optimum asphalt binder content is acceptable. Otherwise, the mixture should be redesigned.

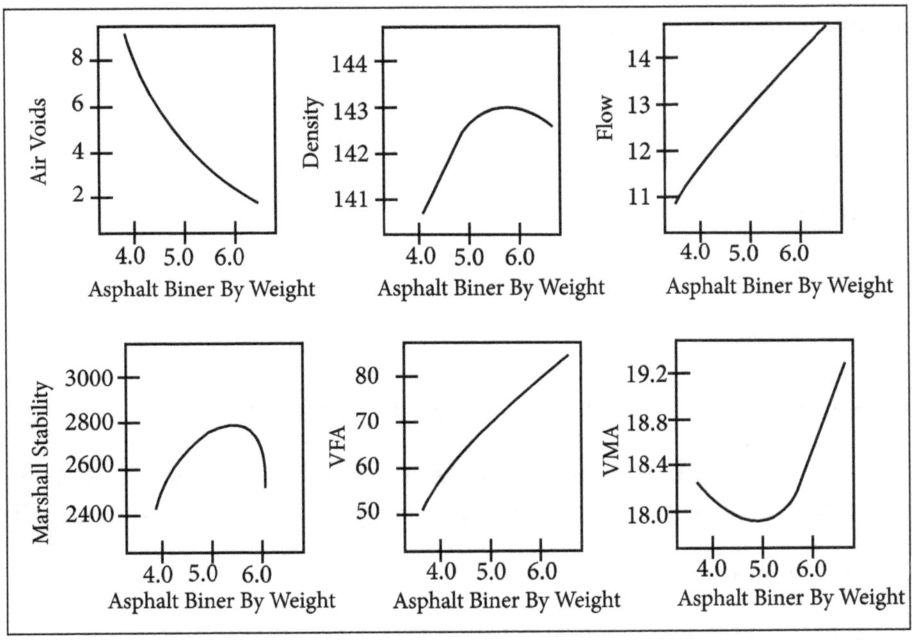

Design graphs for Marshall mix designs.

5

Design of Pavements

5.1 Types of Pavements

Pavements components are as follows:

1. Soil
2. Sub-grade
3. Sub-base
4. Base course
5. Surface course

Requirements of Different Pavement Components

Selection of pavement components for the selected pavements such as type of pavement, base course, sub-base and surface course is governed by the following factors:

1. Type of subgrade soil
2. Drainage conditions
3. Type of traffic and volume of traffic
4. Availability of construction material along the alignment or within a reasonable distance from the site
5. Availability of plant and equipment
6. Climatic conditions
7. Altitude at which the road is proposed to be constructed
8. Time available to complete the project
9. Availability of funds for construction
10. Fund availability for maintenance

The selection of pavement type depends on the factors mentioned above. Out of them, the important factor is the traffic volume or intensity of traffic.

The type of road which could be considered based on traffic are listed below:

Traffic intensity tonnes per day	Type of road
30 to 200	Low cost
up to 500	Water Bound Macadam
up to 1500	Bituminous
up to 2000	Cement concrete

These roads depend on the thickness of pavement section, type of soil subgrade and the desired service life of the road. Each pavement has its own life span. There should be provision to improve the pavement structure in stages to match with increase in road traffic. In order to accommodate this aspect, stage construction technique should be adopted.

5.1.1 Design Factors

Pavement design comprises of two parts, viz.,

i. Mix design of materials to be used in each component layer

ii. Thickness design of the pavement and the component layer

The mix design is made depending on the type of material to be used for the load and other conditions.

Design Wheel Load

Thickness design of pavement primarily depends upon the design wheel load.

The design wheel load depends on the following:

1. Maximum wheel load
2. Contact pressure
3. Equivalent single wheel load

Maximum Wheel Load

In order to know the way by which load is distributed, the wheel load configuration should be known. Figure shows the typical wheel load configuration of a tractor-trailer unit of a heavy duty vehicle. Before deciding the wheel load configuration to be selected, the legal axis load specified in the area should be taken into account. Indian Roads Congress has specified the maximum legal load as 8170 kg and maximum single wheel load as 4085 kg.

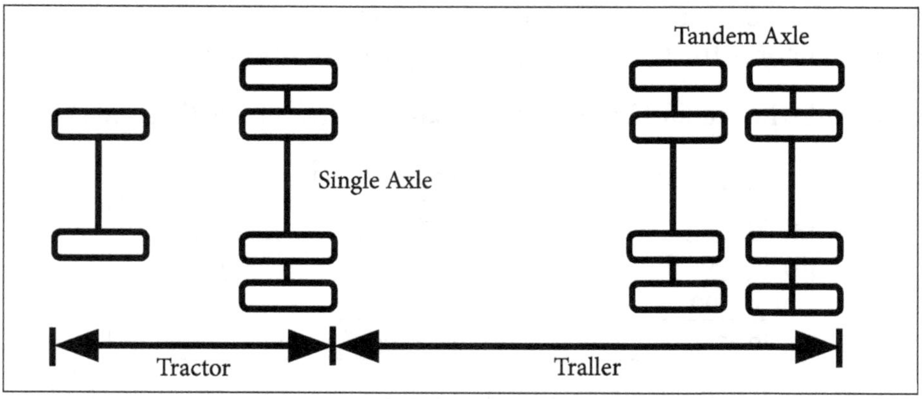

Wheel configuration of tractor-trailer unit.

The configuration or shape of wheel load gives an idea how the load of a vehicle acts on the surface of the pavement. The vertical stress σ^z at depth z, can be computed using Boussinesq's equation under a uniformly distributed circular load as,

$$\sigma_x = p\left[1 - \frac{z^3}{\left(\alpha^2 + z^2\right)^{3/2}}\right]$$

where,

σ^z = Vertical stress at depth Z

p = Surface pressure

Z = Depth at which σ^z is computed

a = Radius of loaded area.

Using the above equation the variation of vertical stress with respect to depth can be computed. The distribution will have the highest value near the surface and decreases with increase in depth.

Contact Pressure

The wheel load is assumed to be distributed over a circular area. Distributed load is the tyre pressure which is differently referred to as inflation pressure or contact pressure.

Tyre pressure and inflation pressure mean exactly the same. Contact pressure is given by the relationship,

Contact pressure, p = L/A

where,

p = Contact pressure

L = Load on wheel

A = Contact area or area of imprint.

The contact pressure is found to be more than tyre pressure when the tyre pressure is less than 7 kg/cm^2.

The contact pressure will be less when the tyre pressure is more than 7 kg/cm^2. The ratio of contact pressure to tyre pressure is defined as rigidity factor.

Rigidity factor is equal to unity when the average tyre pressure is 7 kg/cm^2. This value is higher than unity for low tyre pressure and less than unity for tyre pressure higher than 7 kg/cm^2.

Equivalent Single Wheel Load (ESWL)

To have maximum wheel load, dual wheel assembly is provided to the rear axles of the load vehicles. Because of this, the load due to both the wheels are not to be transferred to the pavement. But there will be overlap pressure after a certain depth. Actual effect is in between a single wheel load and double the load carried by any one wheel. Stress overlap is presented in figure.

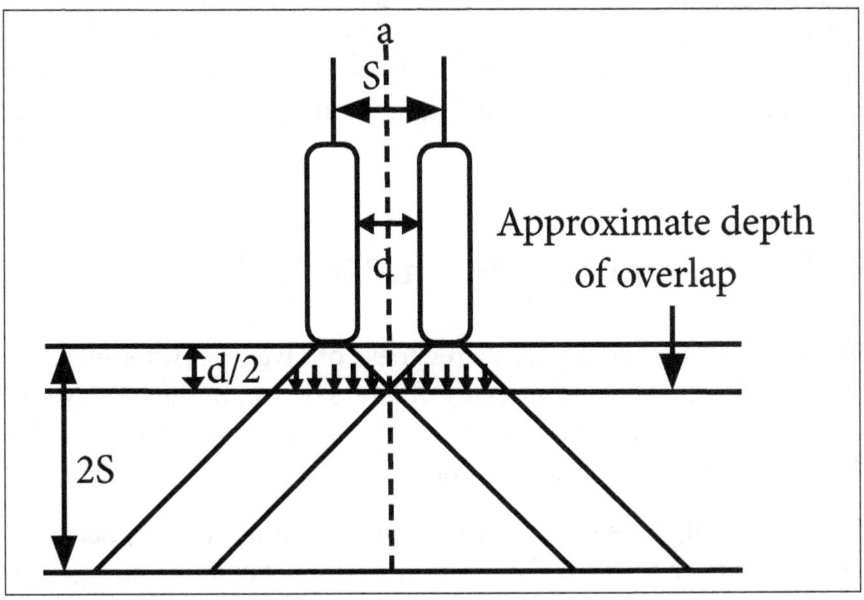

Stress overlap due to dual wheels.

Assume that up to a depth of d/2, the loads act independently beyond which the stresses overlap. The area of overlap becomes more beyond a depth of 2S. Hence consider that the total stress due to the dual wheels at any depth greater than 2S, is equivalent to a single wheel load of 2P magnitude. This stress due to 2P is to be slightly greater than the dual wheel assembly which is on the safe side. This Equivalent Single Wheel Load (ESWL) can be determined by equivalent deflection or equivalent stress criterion.

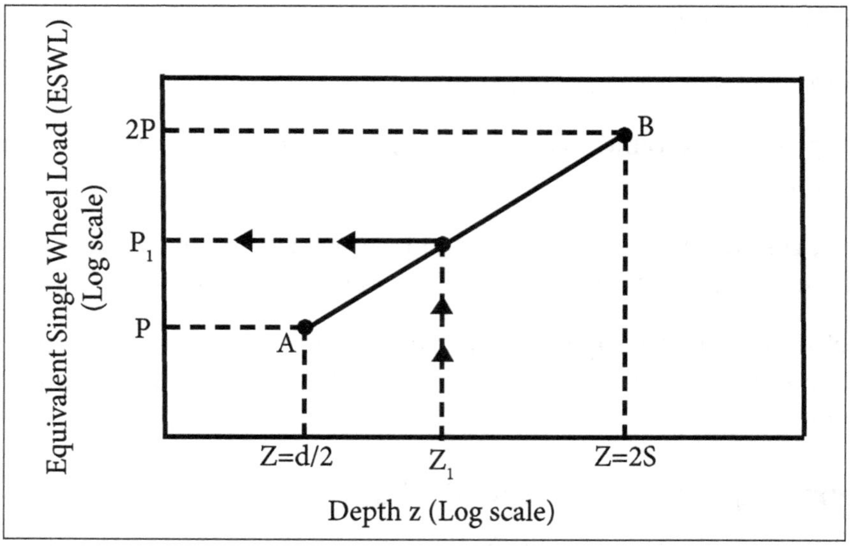

Graphical representation of ESWL.

Line AB represents the locus of points where any single wheel load is equivalent to a certain set of dual wheels. In order to use the graph, for an assumed thickness of pavement an ESWL is got from the graph. This ESWL is used in the design calculations and the thickness of pavement is obtained. If the thickness and assumed thickness are same, then the ESWL assumed is correct. If the design is not repeated and by trial and error, the correct thickness of pavement is obtained. This is valid for the given wheel load configuration. For different wheel load assembly different ESWL plots may be made.

5.2 Flexible Pavements Design Methods

Flexible pavements consist of different thickness of layers of different materials. Due to the wheel loads, the maximum intensity of stress occurs in the top layer of the pavement. Magnitude of stress decreases with depth. Superior quality of materials are used in the top layers followed by less quality materials.

In the design process, it is the basic requirement that under the application of the load none of the layer is over stressed. No section of the pavement undergo excessive deformation which may result in localized depression or settlement. There is no rational design method for flexible pavement. The flexible pavement design methods may be classified for distinct groups:

1. Empirical Methods

These are based on soil classification, physical or strength parameters of soils and other factors such as climate and moisture.

2. Semi-Empirical Methods

These are methods based on stress strain functions, performance and serviceability concepts.

Theoretical Methods

These are based on theoretical analysis and mathematical computations considering stress-strain behaviour of soils.

Various methods which are in use are discussed below:

Group Index Method

This method of flexible pavement design is based on the soil properties like,

1. Particle size distribution
2. Liquid limit
3. Plasticity index.

Group index is claimed to be an inverse measure of the thickness of pavement. The higher is the group index, the lower is the strength and greater is the thickness requirement for a given traffic volume.

According to this method, subgrades have been divided into five based on group index:

Subgrade quality	Value of Group Index
Excellent	0
Good	0 to 1
Fair	2 to 4
Poor	5 to 9
Very poor	10 to 20

Traffic is divided into three groups:

Total volume (commercial vehicles)	Number of vehicles per day
Light	Less than 50
Medium	50 to 300
Heavy	Over 300

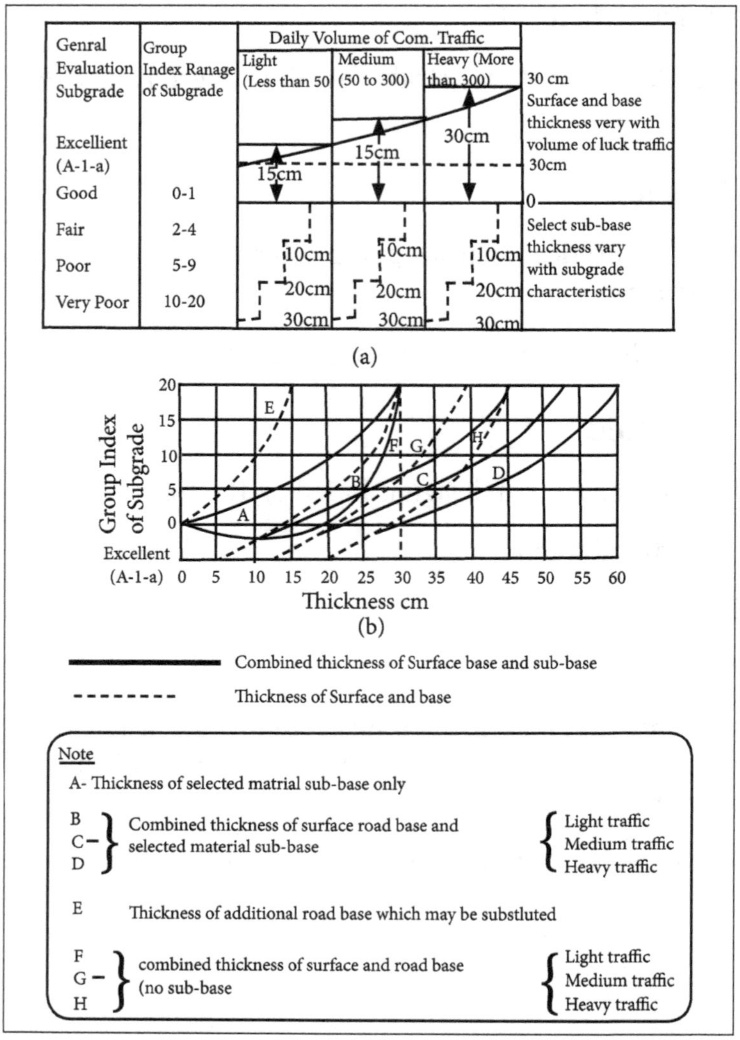

Design chart by group index value.

In order to design the pavement thickness by this method, the following steps are followed:

Step I - The value of group index is found using the details of the soil.

Step II - The anticipated traffic is estimated and classified as light, medium or heavy as indicated in figure.

Step III - The appropriate design curve is chosen from figure and the total thickness of pavement is found for the corresponding group index.

Step IV - The thickness of sub-base or base is found from appropriate curve to find the thickness.

The method is only empirical and considers only the soil physical properties and not the strength and compressibility characteristics of the soil.

5.2.1 CBR Method

California Bearing Ratio (CBR) test is a compressive nature penetration test. This was originally developed by Caltrans that is California Department of Transportation after World War II.

The test is standardized by American Society of Testing Materials (ASTM) as D1883-05 (this standard is used for laboratory prepared samples or re-molded samples) and D4429 (for on field soils); and by American Association of State Highway and Transportation officials (AASHTO) T193; and by British Standard as BS 1377; and by International Standard (IS) as 2720 (Part XVI).

Main Components of Pavement Design

The results obtained by CBR test are used with the empirical curves to determine the thickness of different layers of flexible pavement like subgrade, subbase, base courses. This is the most widely used method for the design of flexible pavement.

CBR value is used to quantify the response of the pavement foundation and the subgrade to loading. CBR does not provide any data regarding the properties of the soil except as to compare its resistance to penetration to the base crushed rock's resistance to penetration. Although CBR test is empirical and has some limitations but it is still used around the world due to its low equipment requirement, ease of performance prediction and history of use.

The two empirical methods for pavement design used now days are CBR method and Group Index Method. In CBR method beside many other steps one step is to determined the CBR value. This test allows the Engineer to design the Capping Layer and the sub-base layer by determining the strength of the underlying soil.

Flowchart for Pavement Design using CBR Method

The Objective of the California Bearing Ratio test is to determine the CBR value for a soil under consideration as a pavement foundation.

CBR is generally used for cohesive soils where effect of water content on CBR is more and generally CBR is determined for a range of water content but it can also be performed for cohesionless soils or coarse grained materials where the effect of water content on CBR is small and thus CBR is performed at Optimum Moisture Content (OMC).

According to ASTM, CBR is primarily intended for evaluating the strength of materials having maximum particle sizes less than ¾ inches or 19 mm, but this restriction is not strict one; if the sample has a material larger than 19 mm; equal amount of such material be replaced by material of size smaller than 19 mm but from the same representative sample.

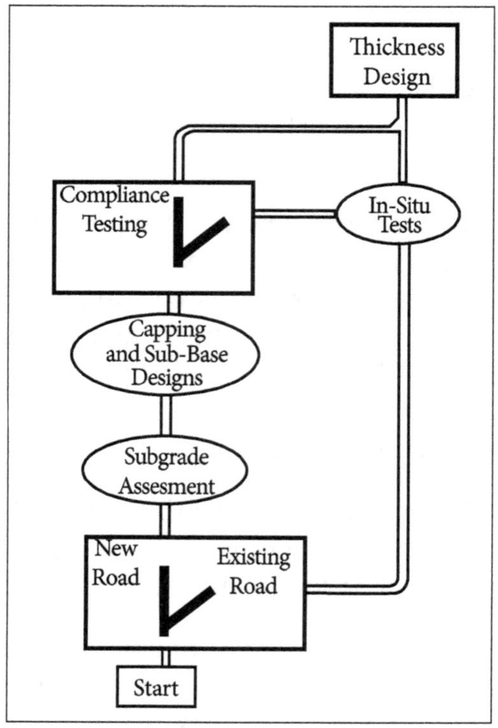

Scope and objective of the CBR Test.

If the sample to be tested contains much fraction of particles of sizes 3 in (Sieve No. 4), the CBR results will fluctuate and thus more trials are required to establish a reliable CBR; thus generally the sample to be tested consists of particles smaller than 19 mm but larger than 3 in.

CBR test can be performed for both in soaked conditions or un-soaked conditions; but mostly soaked conditions is preferred so as to evaluate the material strength in worst conditions.

Description of Apparatus

Apparatus required to perform the CBR test is as follows:

Spacer Disk

A circular disk of metal 5 – 15/16 in. diameter and 2.416 in. height.

Molds

Cylindrical mold with an internal diameter of 6 in. and a height of 7 in. with an extension collar of 2 in. height and a perforated base plate.

Surcharge Weights

Several slotted or split metal plates of 149.2 mm; diameter and 5 lb weight.

Rammer

A rammer of mass 4.54 kg (10 lbs). It is the apparatus for measuring expansions. This consists of a swell plate with adjustable stem and a tripod support for a dial indicator.

Loading Device

A compression type apparatus capable of applying a uniformly increasing load up to 10,000 lb at a rate of 1.3 mm/min.

Penetration Piston

A metal piston of circular cross-section having diameter of 1.954 inches, Area = 3 in^2 and not less than 4 inches in length.

Soaking Tank

A tank suitable for maintaining the water level of 1 inch above the top of specimen.

Drying Oven

Oven capable of maintaining a temperature of $11^\circ + 5^\circ C$ for drying samples. It is a moisture content container.

Miscellaneous

Tools such as mixing pans, spoons, straightedge, filter paper, balances etc.

Test Procedure

Approximately 18 kg soil pass of 19mm sieve and retain of sieve no. 4 is taken.

Moisture and dry density curve is obtained using the standard AASHTO T 99 or T 180.

Optimum Moisture Content (OPC) is obtained from the graph between moisture content and dry density

Prepare the sample by adding optimum moisture content and then compact the soil in five layers by applying 10, 30 and 65 blows respectively in three CBR molds using 10 lb rammer having 18 in. height of fall. The compacted densities of the three specimens range from 95 percent to 100% of the maximum dry density already determined by the T 180 compaction test.

Soaking: Place the swell plate with adjustable stem on the soil sample in the mold and apply sufficient annular weights to produce an intensity of loading equal to the mass of sub-base and base courses and surfacing above the tested material, but not less than 4.54 kg (10 lbs). Place the tripod with dial indicator on top of the mold and make an initial dial reading.

Immerse the mold in water to allow free access of water. Place the sample in water for 96 hours (4 days).

Make a dial reading on soaked specimen and calculate swell as a percentage of initial sample height.

Remove the sample from tank and allow to drain for 15 minutes.

Penetration Test: Place the mold on the loading frame and adjust its potion until the piston is centered on the specimen.

Seat the penetration piston with a 44 N (10 lb) load, and set both the load dial and the strain dial to zero. This initial load is considered as the zero load when determining the stress-penetration relationship.

Place the surcharge weights on the specimens equal to that used during soaking. Apply load at a rate of 1.3 mm/min and record the load for penetration of 0.025 in, 0.05 in, 0.075 in, 0.10 in and so on up to 0.5 inches.

Stress strain curve: Plot curves between load and penetration for each specimen. Apply the corrections to the curves if required. Take the readings of load for 0.1 in and 0.2 in. penetration and find CBR for both penetrations. The greater values is the required CBR for that specimen. Also find the dry density for each specimen.

$$CBR = \frac{\text{Test load value}}{\text{Standard load}} \times 100$$

Design CBR: It is calculated by plotting a graph between CBR values and dry densities of all the three specimens and then calculating the design CBR against value of 85% maximum dry density.

Test Results, Table, Graphs and Calculations

Step 1: Getting the relationship and graph between moisture content and dry density.

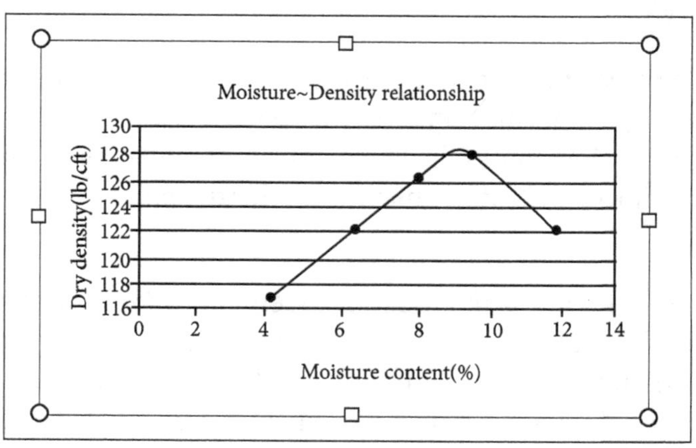

Dry density (lb/cft).

Sample No.	Moisture Content (%)	Dry Density (lb/cft)
1		
2		
3		
4		
5		

Step 2: Finding the density of the three samples each of 10 blows, 30 blows and 60 blows

Step 3: Load v/s Penetration Graph

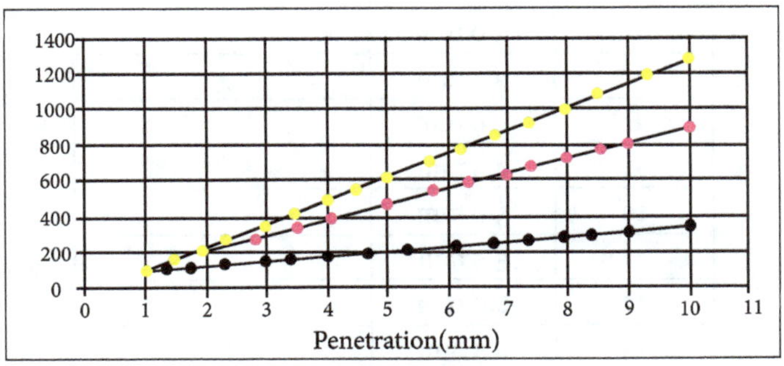

Load penetration curve.

Table: For sample made with 10 blows.

Sr. No	Load (lbs)	Penetration (mm)
1	0.5 mm	
2	1.0 mm	
3	1.5 mm	
4	2.0 mm	
5	2.5 mm	
6	3.0 mm	
7	3.5 mm	
8	4.0 mm	
9	4.5 mm	
10	5.0 mm	
11	5.5 mm	
12	6.0 mm	
13	6.5 mm	
14	7.0 mm	
15	7.5 mm	
16	8.0 mm	

17	8.5 mm	
18	9.0 mm	
19	9.5 mm	
20	10 mm	

Table: For sample made with 30 blows.

Sr. No	Load (lbs)	Penetration (mm)
1	0.5 mm	
2	1.0 mm	
3	1.5 mm	
4	2.0 mm	
5	2.5 mm	
6	3.0 mm	
7	3.5 mm	
8	4.0 mm	
9	4.5 mm	
10	5.0 mm	
11	5.5 mm	
12	6.0 mm	
13	6.5 mm	
14	7.0 mm	
15	7.5 mm	
16	8.0 mm	
17	8.5 mm	
18	9.0 mm	
19	9.5 mm	
20	10 mm	

Table: For sample made with 60 blows.

Sr. No	Load (lbs)	Penetration (mm)
1	0.5 mm	
2	1.0 mm	
3	1.5 mm	
4	2.0 mm	
5	2.5 mm	
6	3.0 mm	
7	3.5 mm	
8	4.0 mm	
9	4.5 mm	
10	5.0 mm	

11	5.5 mm	
12	6.0 mm	
13	6.5 mm	
14	7.0 mm	
15	7.5 mm	
16	8.0 mm	
17	8.5 mm	
18	9.0 mm	
19	9.5 mm	
20	10 mm	

Limitations

The laboratory and field compaction methods are not identical, however, comparative tests indicate that reasonable correlation of results can be obtained from field compact materials and samples compacted under similar conditions in the laboratory.

Because added strength to highly stabilized surfaces such as asphaltic concrete is neglected, the assumption of a completely saturated subgrade condition sometimes results in a too conservative factor of safety.

5.2.2 IRC Method

Recommendations of Indian Roads Congress

Indian Roads Congress (IRC:37-1970) has recommended some important aspects to be considered while using the design chart. Following are the recommendations:

i. The specimen to be tested for CBR should be remoulded specimen prepared preferably by static compaction wherever possible or by dynamic compaction. The standard test procedure should be strictly followed.

ii. In-situ test specimens are not recommended.

iii. For new roads the subgrade soil specimen should be compacted to proctor density at OMC. If the compaction equipment is not available in the field, the specimen may be compacted to the expected field density.

iv. For the existing roads the specimen should be compacted to field density of subgrade soil at water content equal to OMC or field moisture content.

v. For all new constructions the specimen should be soaked for four days prior to testing. This condition is not mandatory for arid climatic regions or regions with annual rainfall less than 50 cm or the water table is very deep or when thickness of impermeable bituminous surfacing is provided.

vi. Atleast three specimens should be tested with identical specimens. If the variation is maximum beyond the norms, then average of six specimens CBR values should be taken. The specified limits of maximum variation in CBR values are:

3% for CBR value up to 10%

5% for CBR value 10 to 30%

10% for CBR value 30 to 60%

vii. The top 50 cm of subgrade should be compacted with density equal to 95 to 100% of proctor density.

viii. Keeping in view the existing traffic and the anticipated growth in traffic should be calculated for atleast 10 years of life period.

The following formula may be used:

$$A = P[l+r]^{n+10}$$

where,

A = Number of heavy vehicles per day for design (laden weight > 3 tonnes)

P = Number of heavy vehicles per day at last count

r = Annual rate of increase of heavy vehicles

n = Number of years between the last count and the year of completion of construction

The value of P has to be found for seven-day heavy vehicles obtained from 24-hours count.

If a reasonable value of r is not available a value of 7.5% may be assumed for rural roads.

ix. The traffic obtained from the above equation has to be used in choosing the appropriate design curve (A to G).

x. The design thickness corresponding to a single axle load up to 8200 kg and tandem axle loads up to 14500 kg is adopted.

xi. Substandard sub-bases with substantial proportion of aggregates of size above 20 mm should not be used in design.

xii. Thin layers of wearing course such as surface dressing or open graded premixed carpet up to 2.5 cm thickness should not be counted towards total thickness as these materials do not contribute to the structural capacity of the pavement.

Design Criteria

The flexible pavements has been modeled as a three layer structure and stresses and strains at critical locations have been computed using the linear elastic model. To give proper consideration to the aspects of performance, the following three types of pavement distress resulting from repeated (cyclic) application of the traffic loads are considered:

- Horizontal tensile stress or strain at the bottom of the bituminous layer which can cause fracture of the bituminous layer.
- Vertical compressive strain at the top of the sub-grade which can cause subgrade deformation resulting in permanent deformation at pavement surface.
- Pavement deformation within the bituminous layer.

While the permanent deformation within the bituminous layer may be controlled by meeting the mix design requirements, thickness of granular and bituminous layers are selected using analytical design approach so that strains at the critical points are within the allowable limits. For calculating the tensile strains at the bottom of the bituminous layer, the stiffness of dense bituminous macadam (DBM) layer with a 60/70 bitumen has been used in the analysis.

Design Procedure

Based on the performance of existing designs and using the analytical approach, simple design charts and a catalogue of pavement designs are added in the code. The pavement designs are given for the subgrade CBR values ranging from 2% to 10% and design traffic ranging from 1 msa to 150 msa for an average annual pavement temperature of 35°C. The later thicknesses obtained from the analysis has been slightly modified to adapt the designs to stage construction.

Using the following simple input parameters, appropriate designs can be chosen for the given traffic and soil strength:

- Design traffic in terms of cumulative number of standard axles
- CBR value of subgrade

Design Traffic

The method considers traffic in terms of the cumulative number of the standard axles (8160 kg) to be carried by the pavement during the design life. This requires the following information:

1. Distribution of commercial traffic over the carriageway

2. Design life in number of years
3. Initial traffic in terms of CVPD
4. Traffic growth rate during the design life
5. Vehicle damage factor (VDF)

Pavement Thickness Design Charts

For design of pavements to carry traffic in the range of 1 to 10 msa, use chart 1 and for traffic in the range of 10 to 150 msa, use chart 2 of IRC:37 2001. The design curves relate pavement thickness to the cumulative number of standard axles to be carried over design life for different sub-grade CBR values ranging from 2% to 10%. The design charts will give total thickness of the pavement for the above inputs. The total thickness consists of bituminous surfacing, granular sub-base, and granular base. The individual layers are designed based on the recommendations given below and subsequent tables.

Pavement Composition

Sub-base

Sub-base materials comprise brick metal, natural sand, gravel, laterite, crushed stone or combinations thereof meeting the prescribed grading and physical requirements. The sub-base material should have a CBR of atleast 20% and 30% for traffic upto 2 msa and traffic exceeding 2 msa respectively. Sub-base generally consist of granular or WBM and the thickness should not be less than 150 mm for design traffic less than 10 msa and 200 mm for the design traffic of 1:0 msa and above.

Base

The recommended designs are for unbounded granular bases that comprise conventional water bound macadam (WBM) or wet mix macadam (WMM) or equivalent confirming to MOST specifications. The materials must be of good quality with minimum thickness of 225 mm for traffic up to 2 msa an 150 mm for the traffic exceeding 2 msa.

Bituminous Surfacing

The surfacing consists of a wearing course or a binder course plus wearing course. The commonly used wearing courses are surface dressing, mix seal surfacing, semi-dense bituminous concrete, open graded premix carpet, and bituminous concrete. For binder course, MOST specifies, it is desirable to use bituminous macadam (BM) for traffic upto 0.5 msa and the dense bituminous macadam (DBM) for traffic more than 5 msa.

5.2.3 Burmister Method

The first attempt to calculate displacements due to loading on an elastic half-space, like the surface of an homogeneous material with an infinite area and depth, was made by Kelvin in 1868 (Croney and Croney, 1997). Later, the Boussinesq's solution (1885) for a concentrated load became a fundamental tool to compute stress, strain and deflection. The solution can be integrated to obtain responses due to a general surface load, including a circular loaded area (Huang, 1993).

The concept of the multi-layer analysis has its roots in the Burmister two-layer and three-layer solutions (Burmister, 1945); charts and the tables summarizing these solutions were developed later (Foster and Alvin, 1954; Burmister, 1958; Jones, 1962; Huang, 1969 and 1973).

Burmister's layered theory can be applied to a multi-layer system of the linear elastic materials structured on top of a half space subgrade following the basic assumptions (Huang, 1994):

- The material is weightless and horizontally infinite.
- Each layer is homogeneous, isotropic, and linearly elastic, characterized by the Young's modulus of elasticity, E, and Poisson's ratio, ν.
- The load is uniformly applied on the surface over a circular area.
- The thickness of each layer is finite, and the subgrade is considered as infinite layer.
- Continuity conditions are satisfied at layer interfaces.

In the M-E PDG, the MLET is implemented in a modified version of the JULEA algorithm (NCHRP, 2004). By the principle of superposition, single wheels can be combined spatially into multi-wheel axles to simulate various axle configurations.

The small set of input parameters required by MLET facilitates its implementation and use. The inputs that are required are the layer thicknesses, the elastic properties for each layer, the tire pressure, and tire contact area. The main disadvantage of MLET is its inability to consider nonlinearities often exhibited by pavement materials.

5.2.4 Mechanistic Method

Mechanics is the science of motion and the action of forces on bodies. A mechanistic approach seeks to explain this phenomena only by reference to physical causes.

In pavement design, the phenomena are the strains, stresses and deflections within a pavement structure, and the physical causes are the loads and the material properties of the pavement structure.

Mechanistic Design

A method that involve numerical capability to calculate the stress, strain, or deflection in a multi-layered system, like a pavement, when subjected to external loads, or the effects of temperature or moisture.

A method that cite the ability to translate the analytical calculations of pavement response to performance.

Mechanistic design procedure are based on assumption that a pavement can be modeled as multilayered elastic or visco-elastic structure on an elastic or visco-elastic foundation.

Benefits

- Improved reliability for design.
- Ability to predict specific types of distress.
- Damaging effects of high tire pressure, increased loads, multiple axles can be modeled.
- Ability to extrapolate from limited field and laboratory results.
- Better utilization of available materials.
- Improved method for premature distress analysis.

Empirical-Mechanistic Method

Empirical-Mechanistic method of design is based on the mechanics of materials that relates input, like wheel load, to an output or pavement response. In pavement design, the responses are the deflections, stresses and strains within a pavement structure and the physical causes are the loads and material properties of pavement structure.

The relationship between these phenomena and their physical causes are typically explained using some mathematical models. Along with this mechanistic approach, empirical elements are used while defining the value of the calculated stresses, strains, and deflections that result in the pavement failure.

The relationship between the physical phenomena and pavement failure is described by empirically derived equations that compute the number of loading cycles to failure.

EICM Environmental Model

EICM is a mechanistic model of one dimensional heat and the moisture flow that simulates the changes in the behavior and characteristics of the pavement and subgrade materials induced by the environmental factors (NCHRP, 2004). Daily and seasonal variations of the temperature and moisture within the pavement structure are induced by weather history at the project site.

Various material types have different responses to climatic variations. The unbound materials are affected by moisture change and by freeze-thaw cycles during winter and spring seasons. Asphalt concrete responds to the temperature variations, which affects directly the dynamic modulus of the mixture. Temperature is also cause of the thermal cracks, either from a single thermal variation or from repetitive cycles of cool/warm temperatures.

The EICM consists of three major components:

1. The Climatic-Materials-Structural Model (CMS Model) was originally developed at the University of Illinois (Dempsey et al., 1985).
2. The CRREL Frost Heave and Thaw Settlement Model (CRREL Model) was originally developed at the United States Army Cold Regions Research and Engineering Laboratory (CRREL) (Guymon et al., 1986).
3. The Infiltration and Drainage Model (ID Model) was originally developed at Texas A&M University (Lytton et al., 1990).

In case of the flexible pavements, three major environmental effects are of particular interest:

- Moisture variation for subgrade and unbound materials. The resilient modulus input of unbound materials is defined as being at optimum density and moisture content. A correction factor is defined to adjust resilient modulus based on predicted moisture content.
- Temperature variations for the asphalt concrete. Dynamic modulus of asphalt concrete mixtures is very sensitive to temperature. Temperature distributions in the asphalt concrete layers are predicted and then used to define the stiffness of the mixture throughout the sublayers. Temperature distributions are also used as inputs for the thermal cracking prediction model.
- Freezing and thawing for the subgrade and unbound materials. The resilient modulus of unbound materials located within the freezing zone increases during freezing periods and decrease during thawing periods. The EICM predicts formation of the ice lenses and defines the freezing zone.

Failure Criteria

The main empirical portions of mechanistic-empirical design process are the equations used to compute the number of loading cycles to the failure. These equations are derived by observing the performance of the pavements and relating the type and extent of the observed failure to an initial strain under various loads. Currently, two types of failure criteria are widely recognized, one relating to the fatigue cracking and the other to rutting initiating in the subgrade.

5.2.5 IRC Method for Low Volume Flexible Pavements

Low volume roads are divided into three categories such as:

1. Gravel/Aggregate Surfaced (Un-paved roads)
2. Flexible Pavements (Paved roads)
3. Rigid Pavements

Flexible pavements have a bituminous, aggregate, or soil surface on which the vehicles travel. If the surface is asphalt, there may be more than one layer, with the surface course or wearing course underlain by one or more asphalt layers called the binder(s) or base course(s). The surface course may be a bituminous surface treatment ("tar and chip," "chip seal"), or it could be a granular layer consisting of crushed rock, crushed gravel, or gravel. Some designers of industrial haul roads call the surfacing course the capping layer.

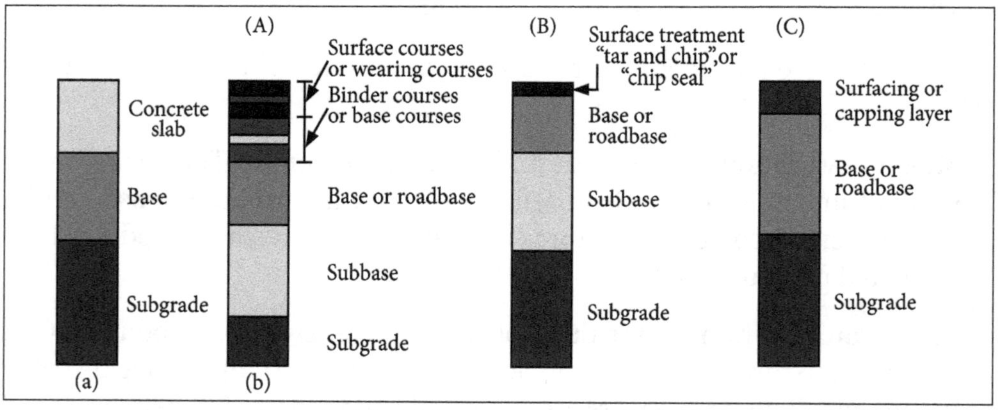

Pavement layers in (a) rigid pavements, and (b) flexible pavements [(A) asphalt, (B) surface-treated, (C) aggregate].

Beneath the surface layer, flexible pavements usually have a granular base or road base. The base layer is usually designed to be well draining, so that water that penetrates the surface layer can be carried to the sides, out of the pavement structure. Water movement is enhanced by having all layers in the pavement crowned (cambered or constructed with a crossfall). Crossfalls of 2% and 4% are appropriate for asphalt and granular road surfaces, respectively. The granular road surfaces need a higher crossfall to overcome the greater impedance presented by the rough surface to the flow of water, because rutting may counteract drainage, and because grading operations tend to flatten the crossfall. It is a common error to specify too little crossfall for granular road surfaces-2% is inadequate.

Under the base layer the pavement may be designed with a subbase layer. The subbase is usually gravelly or sandy well-draining pit-run material. Beside drainage and protection of the subgrade beneath it, in cold regions the subbase is constructed of non-frost-susceptible material. This serves to increase the depth of non-frost-susceptible materials in the pavement structure, so the penetration of the frost front does not reach frost-susceptible soil and cause heaving.

The subbase is in turn underlain by the subgrade, the native soil at the site (or the top of embankment fill if it exists). During construction, the subgrade should be proof-rolled before the pavement structure is built on it, and any soft, loose, wet, organic, or otherwise weak material sub-excavated and replaced with clean, well-compacted subbase material fill.

Variations

There are numerous variations beyond those listed here. For example, asphalt variations include full depth and deep strength pavements. The granular layers of any of the types of pavements can be treated with Portland cement, lime, fly ash, bitumen, or other chemicals or additives (as appropriate to the material) to stiffen them. Composite pavements are rigid pavements with slab surfaced with asphalt. "White-topping" is the inverse of that, with Portland cement placed on an asphalt pavement.

5.3 Rigid Pavements: Design Considerations

The distance from the intersection where this deformation like rutting and shoving occurs varies depending on traffic situation, types of traffic, speed and stopping distance, and number of vehicles per lane stopped at the intersection. Various approaches can be used.

In case of some applications, PCCP can extend full width for several hundred feet on each side of the intersection. In other situations, concrete lanes approaching the intersection extend 250 feet, while those going away terminate about 60 feet beyond the curb return. These approaches can be used for both high volume streets and bus stops. For more moderate traffic, around 50 to 100 feet on each side of the intersection is likely to be sufficient. This distance can be based on an evaluation of existing traffic and pavement conditions.

Dowels should be placed in the transverse joints of dominant traffic stream, as well as, the cross street transverse joints. Tie bars must be placed in the longitudinal joints of dominant traffic stream and cross street sections past the intersection.

Class P concrete is recommended for rigid pavements. If it is recommendable to fast track an intersection reconstruction, the Class E concrete can be used. Class E concrete is designed to achieve a minimum of 2,500 psi in 12 hours or as required. It is possible to remove existing pavement, recondition base materials, place Class E concrete and have roadway open for traffic in 24 hours.

5.3.1 Wheel load Stresses

The cement concrete slab is assumed to be homogeneous and to have a uniform elastic properties with vertical sub-grade reaction being proportional to the deflection. Westergaard developed relationships for the stress at interior, edge and corner regions, which is denoted as σ_i, σ_e, σ_c in kg/cm² respectively and given by the following equation,

$$\sigma_i = \frac{0.316P}{h^2}\left[4\log_{10}\left(\frac{l}{b}\right)+1.069\right]$$

$$\sigma_e = \frac{0.572P}{h^2}\left[4\log_{10}\left(\frac{l}{b}\right)+0.359\right]$$

$$\sigma_c = \frac{3P}{h^2}\left[1-\left(\frac{\alpha\sqrt{2}}{l}\right)^{0.6}\right]$$

where,

h is the slab thickness in cm.

P is wheel load in kg.

a is the radius of the wheel load distribution in cm.

l is the radius of the relative stiffness in cm.

b is the radius of the resisting section in cm.

5.3.2 Temperature Stresses

Temperature stresses are developed in cement concrete pavement due to variation in slab temperature. This is caused by:

i. Daily variation that results in a temperature gradient across the thickness of the slab.

ii. Seasonal variation resulting in overall change in slab temperature.

The former results in warping stresses and the later in frictional stresses.

Warping Stress

The warping stress at interior, edge and corner regions, denoted as σ_{ti}, σ_{te}, σ_{tc} in kg/cm² respectively and given by,

$$\sigma_{ti} = \frac{E\in t}{2}\left(\frac{C_x + \mu C_y}{1-\mu^2}\right)$$

$$\sigma_{te} = \text{Max}\left(\frac{C_x E\in t}{2}, \frac{C_y E\in t}{2}\right)$$

$$\sigma_{tc} = \frac{E\in t}{3(1-\mu)}\sqrt{\frac{\alpha}{l}}$$

where,

E is the modulus of elasticity of concrete in kg/cm².

$\in$ is the thermal coefficient of concrete per °C.

t is the temperature difference between the top and bottom of the slab.

C_x and C_y are the coefficient based on L_x/l in the desired direction and L_y/l right angle to the desired direction.

μ is the Poisson's ration (0.15).

a is the radius of the contact area.

L is the radius of the relative stiffness.

5.3.3 Frictional Stresses

The frictional stress σ_f in kg/cm² is given by,

$$\sigma_f = \frac{WLf}{2\times10^4}$$

where,

W is the unit weight of concrete in kg/cm² (2400).

f is the coefficient of sub-grade friction (1.5).

L is the length of the slab in meters.

5.3.4 Combination of Stresses

The cumulative effect of the different stress give rise to the following three critical cases:

Summer, Midday

The critical stress is for edge region is given by,

$$\sigma_{critical} = \sigma_e + \sigma_{te} - \sigma_f$$

Winter, Midday

Critical combination of stress is for the edge region given by,

$$\sigma_{critical} = \sigma_e + \sigma_{te} + \sigma_f$$

Midnights

The critical combination of stress is for corner region is given by,

$$\sigma_{critical} = \sigma_e + \sigma_{te}$$

5.3.5 Design of Slabs

Critical Stress Condition

In-service cement concrete pavements are subjected to stresses due to a variety of factors, acting simultaneously. The severest combination of various factors that induce the maximum stress in the pavement will give the critical stress condition. The factors commonly considered for the design of pavement thickness are: flexural stresses due to traffic loads and the temperature differentials between the top and bottom fibres of the concrete slab, as the two are assumed to be additive under critical condition. The effects of moisture changes are opposite of those of temperature changes and are, not generally considered critical to thickness design.

The loads applied by single as well as tandem axles cause maximum flexural stresses when tyre imprint touches the longitudinal edge as shown in figure. When the tyre imprints touch the transverse joints with or without the dowel bar, part of the load is transferred to the other side of the slab by the aggregate inter-lock or dowel bar causing lower flexural stress both along the corner as well as along the transverse joint.

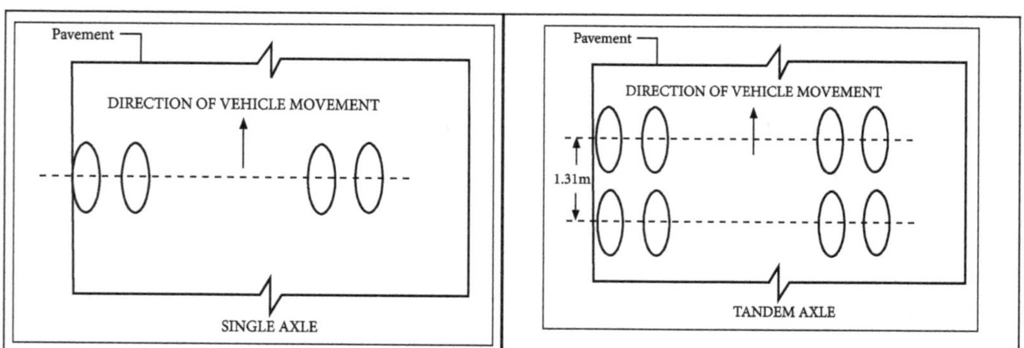

Lateral placement of wheel.

In case the slab is cast panel by panel with a clear vertical break without any dowel bar or an aggregate inter-lock, corner load stresses are critical when dual wheel system is at the corner. The tandem axles carrying twice the load of a single axle cause flexural stresses which are about 20 per cent lower than that of single axle load because of superposition of negative bending moment due to one dual wheel load over the other. The average spacing of the tandem axle is taken as 1.31 metres. Tandem and tridem axle loads may cause loss of subgrade because of higher deflection.

In such case, the additional design criterion of erosion can be included based on experience. Cement concrete pavements undergo a daily cyclic change of the temperature differentials, the top surface being hotter than the bottom during the day, and cooler during night. Consequent tendency of the pavement slabs is to warp upwards during the day and downwards during the night. The restraint offered to this warping tendency by self-weight and the dowel bars of the pavement induces stresses in the pavement, referred to commonly as temperature warping stresses. The warping stresses are flexural in nature, being tensile at the bottom during the day and at top during the night.

As restraint offered to warping at any section of slab would be a function of weight of the slab upto that section, it is obvious that the corners have very little of such restraint for slabs without dowel bars. The restraint is high in the slab interior and somewhat less at the edge. As a result, the temperature stresses induced in pavement are maximum at the interior. Under the action of the load applications, maximum stress is induced in the corner region if the joints are not provided with dowel bars, as the corner is discontinuous in two directions. The corner tends to bend like a cantilever, producing tension at the top during night hours, whereas, tension is produced during the daytime at the bottom of the slab in the interior as well as at edge.

The maximum combined tensile stress in the three regions of the slab will thus be caused when the effects of temperature differentials are such as to be additive to load effects. This would occur during the day in the case of the interior and edge regions at time of maximum temperature differential in the slab. In the corner region, temperature stress is negligible but load stress is maximum at night when the slab corners have a tendency to lift up, due to the warping and lose partly the foundation support.

Considering the total combined stress for the three regions, viz., corner, edge and interior, for which load stress decreases in that order while the temperature stress increases, critical stress condition is reached in the edge region. It is, therefore, necessary that the concrete slab is designed to withstand stresses due to warping and the wheel load at the edge region. It is also necessary to check the stress at the corner region if dowel bars are not provided at the transverse joints and if there is no possibility of the load transfer by aggregate inter-lock.

Calculation of Stress

Edge Stress

a. Due to Load

Since loads causing failure of pavements are mostly applied by single and tandem axles, stresses must be determined for the condition shown in figure. Picket & Ray's chart can be used for the stress computation in the interior as well as at the edge. By the fundamental concept of Westergaard and Picket & Ray's pioneering work, computer programme IITRIGID developed at IIT Kharagpur was used for the computation of stress for the edge dad condition shown in figure.

In the earlier version of IRC:58-1988, the calculation of load stresses was done as per Westergaard's equations which was modified by Teller and Sutherland. The use of these equations has its own limitations because the configuration of the wheels are neglected. Though, these equations give stresses which are not very much in variance with stresses computed by the programme IITRIGID, it is commended that the stresses calculated from the programme IITRIGID be used in design.

b. Due to Temperature

The temperature stress at the critical edge region may be obtained as per Westergaard's analysis using Bradbury's Coefficient from the following equation.

$$S_{te} = \frac{E\alpha tC}{2}$$

where,

S_{te} = Temperature stress in the edge region, kg/cm^2

E = Modulus of elasticity of concrete, kg/cm^2

t = Maximum temperature differential during day between top and bottom of the slab, °C

α = Coefficient of thermal expansion of cement concrete, per °C

C = Bradbury's coefficient, which can be ascertained directly from Bradbury's chart against values of L/I and B/I

L = Slab length, or spacing between consecutive contraction joints, cm

W = Slab width, or spacing between longitudinal joints, cm and

I = Radius of relative stiffness, cm

$$= \sqrt[4]{\frac{Eh^3}{12\left(1-\mu^2\right)k}}$$

μ = Poisson's ratio

h = Thickness of the concrete slab, cm

k = Modulus of subgrade reaction, kg/cm^3

The values of Bradbury's coefficient C are presented in the form of chart in figure.

Corner Stress

The load stress in the corner region may be obtained as per Westergaard's analysis, modified by Kelly, from the following equation.

$$S_c = \frac{3P}{h^2}\left[1 - \left(\frac{a\sqrt{2}}{1}\right)^{1.2}\right]$$

where,

S_c = Load stress in the corner region, other notations remaining the same as in the case of edge load stress formula, kg/cm²

P = Wheel Load, kg

a = Radius of equivalent circular contact area, cm

The temperature stress in the corner region is negligible, as the corners are relatively free to warp, therefore, may be ignored.

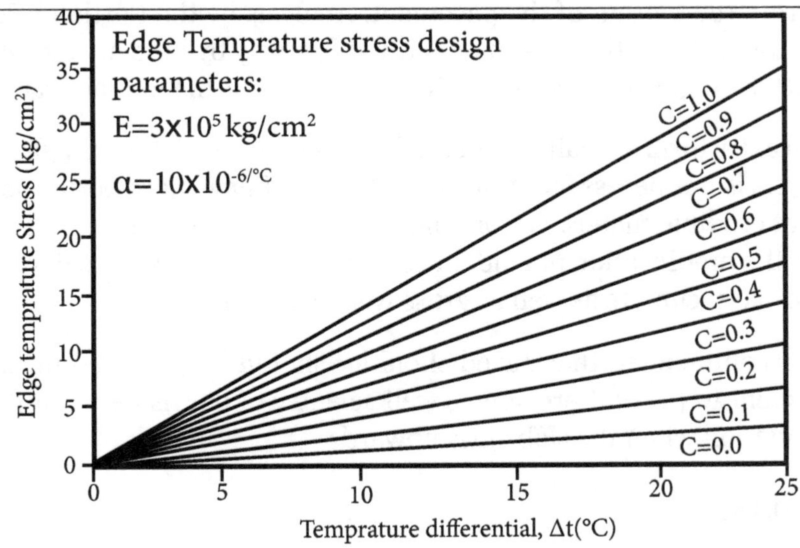

Chart for Determination of Coefficient C:

L/1 or B/1	c	L/1 or B/1	c
1	0.000	7	1.030
2	0.040	8	1.077
3	0.175	9	1.080
4	0.440	10	1.075
5	0.720	11	1.050
6	0.920	12	1.000

Design chart for calculation of edge temperature stress.

Design Charts

A user friendly computer programme is also enclosed in a floppy for computation of stresses at the edges.

Stress Ratio and Fatigue Analysis

For a given slab thickness and other design parameters, the flexural stress at the edge due to the application of a single or tandem axle loads may be determined by using the appropriate stress chart. This stress value is divided by the design flexural strength of the cement concrete, to obtain the stress ratio in pavement. If the stress ratio is less than 0.45, allowable number of repetitions of the axle load is infinity. The cumulative fatigue damage is determined for the different axle loads and the value of the damage should be equal to or less than one.

Erosion Consideration

AASHTO Road Test have indicated that there is an important mode of distress in addition to the fatigue cracking that must be considered in the design. This is the erosion of material from the bottom of the pavement. Analysis by the Portland Cement Association has indicated that the erosion was caused largely by tandem and the multi-axle vehicles and that single axles were mostly responsible for the fatigue cracking.

Since tandem axles form a small part of total commercial vehicles on highways in India, erosion analysis is not necessary at present. Record of pavement performance data including loss of erodible materials from the sub-base of the concrete pavements will be necessary for the modification of guidelines in future since erosion is dependent on the quality of sub-base, climate as well as gross weight of vehicles.

It is further recommended that paved shoulder should be provided up to 1.5 metres beyond pavement to prevent erosion as well as entry of debris between the pavement slab and foundation when the slab curls upwards.

Hard Shoulder

In order to protect the foundation layers from loss of strength due to erosion, a number of measures are taken. Usually, dry lean concrete (DLC) sub-base is extended by 40 to 50 cm towards the shoulder. Additionally, full depth bituminous shoulder or the tied cement concrete shoulder is constructed to protect the pavement edge. The widening rigid pavement to act as a shoulder has also been attempted. With such a shoulder, the load stresses at edges will reduce marginally.

Composite Rigid Pavement

Where the polythene separation layer between the concrete slabs and dry lean concrete

(DLC) sub-base is eliminated a monolithic action of two layers results and this action can be exploited to reduce pavement thickness. The layer below DLC has to be smooth and may warrant an antifriction layer to allow the thermal movements to take place without any hindrance. The appropriate design procedure can be established only on the basis of extensive research.

Anchor Beam and Terminal Slab

During the hot season, the concrete slabs expand and this will result in build-up of horizontal thrust on din-wall/abutment. To contain this thrust, RCC anchor beams are generally provided in the terminal slab. The terminal slab, therefore, will have to be reinforced to strengthen it. The details of the anchor beam and terminal slab are discussed in IRC:15-2002.

Recommended Design Procedure

Step 1: Stipulate design values for the various parameters.

Step 2: Decide types and spacing between joints.

Step 3: Select a trial design thickness of pavement slab.

Step 4: Compute the repetitions of axle loads of different magnitudes during the design period.

Step 5: Calculate the stresses due to single and tandem axle loads and determine the cumulative fatigue damage (CFD).

Step 6: If the CFD is more than 1.0, select a higher thickness and repeat the steps 1 to 5.

Step 7: Compute the temperature stress at the edge and if sum of the temperature stress and the flexural stress due to the highest wheel load is greater than modulus of the rupture, select a higher thickness and repeat the steps 1 to 6.

Step 8 : Design the pavement thickness on the basis of the corner stress if no dowel bars are provided and there is no load transfer due to lack of aggregate inter-lock.

5.3.6 Design of Joints

Joints are used in PCCP to aid the construction and eliminate random cracking. There are two types of longitudinal joints. Longitudinal weakened plane joints relieve the stresses and control longitudinal cracking. Longitudinal construction joints perform same functions and also divide the pavement into suitable paving lanes.

Longitudinal weakened plane joints are spaced to coincide with lane markings, and are formed by sawing the hardened concrete to a depth of 1/3 the pavement thickness. Longitudinal construction joints should be tied with deformed reinforcing steel bars to hold the slabs in the vertical alignment. Stresses in a slab are reduced when the slab is tied to adjacent slabs.

Keyed joints may be used in a longitudinal construction joint, but tying the slabs is preferable. The key may be formed by attaching a keyway at mid-depth of a side form. With a slip-form paver, the keyway can be formed as the paver advances. For detailed layout refer to the CDOT Standard Plans (M & S Standards), July 2012.

Transverse joints are spaced at short intervals in the slab. A maximum of 12 feet is recommended to insure crack control and for ease of construction. The joint should be sawed to a depth of at least 1/3 the pavement thickness.

Dowel bars in the first three transverse joints where the portland cement concrete pavement abuts asphalt pavement can prevent progressive slab movement. Expansion joints are not required except at intersections.

The following summarizes the general design guides and information for constructing the rigid pavement joints:

Joints are used in PCCP to aid construction and minimize random cracking. The odd shaped slabs and acute angles of less than 60 degrees should be avoided. Longitudinal joint spacing should be approximately 12 feet.

Transverse joint spacing must be at regular 12 foot intervals with no more than a 15 foot spacing. The thinner slabs tend to crack at closer intervals than thicker slabs and long narrow slabs tend to crack more than square slabs. All contraction joints should be continuous through the curb and have a depth equal to 1/3 of the pavement thickness.

Transverse joints must be carried through the curb. Expansion joint filler must be full-depth and extend through the curb. Longitudinal joints must be tied to hold adjacent slabs in vertical alignment, as well as, curb and gutter.

The normal backfill behind the curb constrains the slabs and holds them together. Offsets at radius points must be at least 18 inches wide. Minor adjustments in joint location made by skewing or shifting to meet inlets and manholes will improve the pavement performance. When the pavement areas have many drainage structures, particularly at intersections, place joints to meet structures, if possible.

Depending on the type of castings, manhole and inlet frames may be boxed out and isolated using expansion joint filler. The frames may be wrapped with expansion joint filler or the frames can be cast rigidly into concrete.

CDOT designs their PCCP using the JPCP (Jointed Plain Concrete Pavement) method. For a detailed illustration, see CDOT's Standard Plans (M & S Standards), July 2006, M-412-1 and as revised.

Following the previous design of a new intersection near Sugarloaf Reservoir, the following steps and points should be followed to design the slabs and location of joints:

Step 1. Draw all edge of pavement lines on a plan view. Plot all utility manholes, catch basins, water valves, etc. on the plan view.

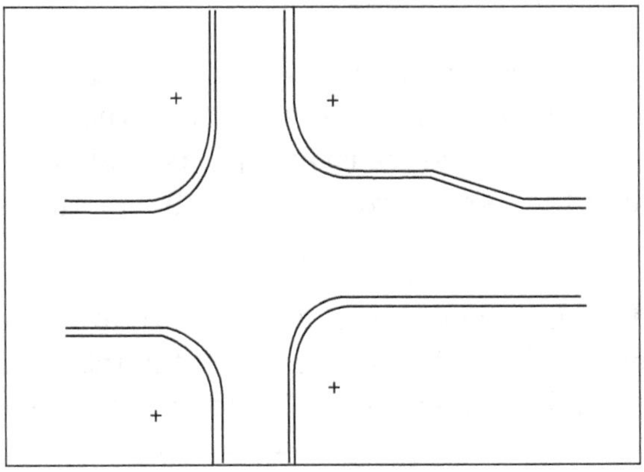

Step 2. Draw lines, which define the median, travel lanes and turning lanes. These lines define the longitudinal joints:

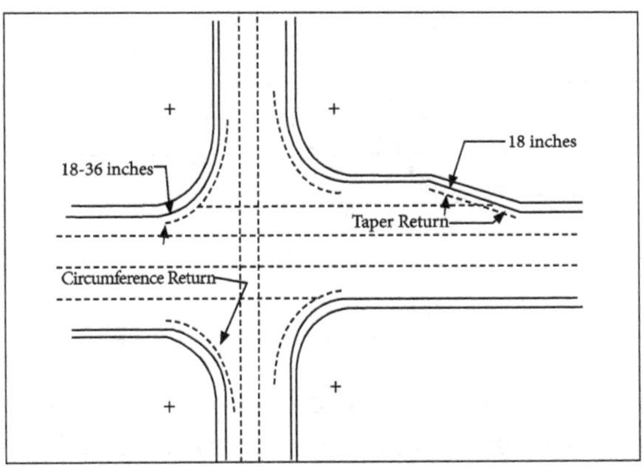

Step 3. Determine locations in which the pavement changes width. Joints at these locations are necessary to isolate irregular shapes. Triangles or circles, which are left intact with a rectangular portion of a slab, will create a plane of weakness which will break during temperature movements of the slab. Concrete simply likes to be square.

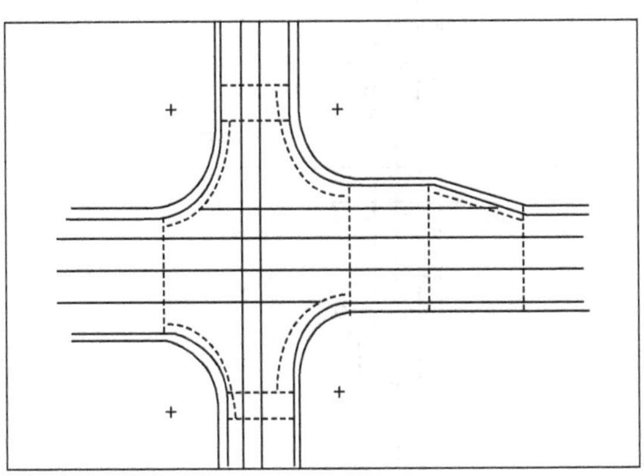

Step 4. Draw transverse lines through each manhole or other utility. Joints need to be placed through the utility structures in the pavement, or movement of the pavement will be restricted and cracking will result.

When the structures are located near a joint placed according to the steps above, isolation can be provided by adjusting the joint to meet the structure. By doing this, numerous short joints will be avoided. Add transverse joints at all locations where pavement changes width, extending the joints through the curb and gutter. Create an "intersection box".

Do not extend joints that intercept a circumference-return or crossroad-return line, except at the tangent points. The joints at the tangent point farthest from mainline becomes an isolation joint in the crossroad for T- and unsymmetrical intersections.

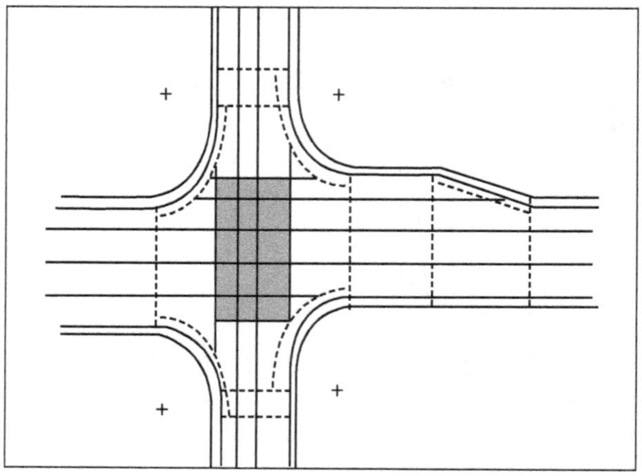

Step 5. The intermediate areas between the transverse joints placed in Steps 3 and 4 may also require transverse joints. These joints are placed by using a standard joint spacing. There is an old rule of thumb for joint placement in plain concrete pavements that says the joint spacing, in feet, should be no greater than two to two and a half times the slab thickness, in inches. Though, in no case should the joint spacing exceed 15 feet.

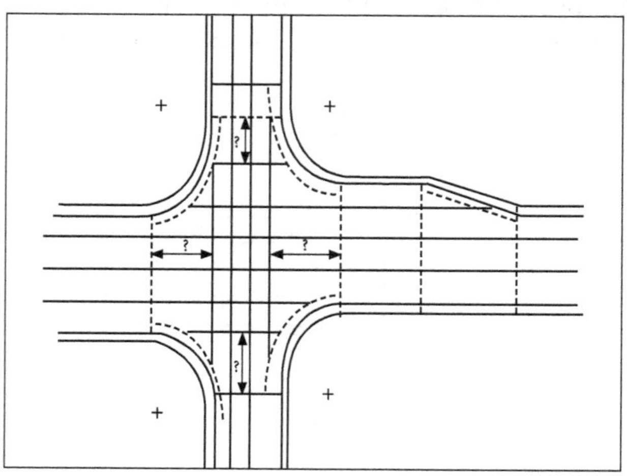

Step 6. Where an intersection is encountered, intermediate joints must be placed. This is done by extending the radius line of each turning radius three feet beyond back of curb. The extension is made at approximately the 45° line for small radii and at approximately the 30° and 60° lines for larger radii. Joints are then connected to each of these points.

Step 7. Expansion joints are needed adjacent to any structure, i.e., bridges, buildings, etc., and at T intersections. T intersections are isolated at radius return to intersecting street. The same layout discussed in the Step 6 is used out to that location.

Step 8. If there are structures such as manholes or any other structures, which cannot be intersected by a joint, they should be isolated. These structures can be isolated by boxing out the structure during paving.

Manholes can also be isolated using a telescoping manhole, that can be poured integral with pavement. The area around the structure must be reinforced to control cracking. The joints that form a box out should be expansion joints to allow some movement.

Step 9. Check the distances between the "intersection box" and surrounding joints.

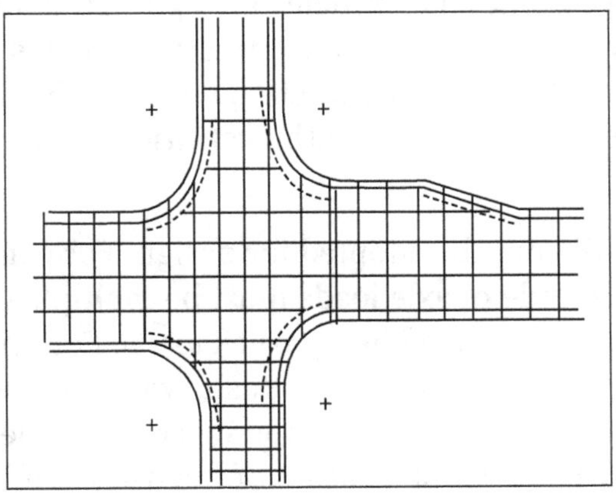

Step 10. Lightly extend lines from the center of the curve(s) to the points defined by "intersection box" and point(s) along any island. Add joints along these radius lines. Finally, make slight adjustments to eliminate dog legs in mainline edges.

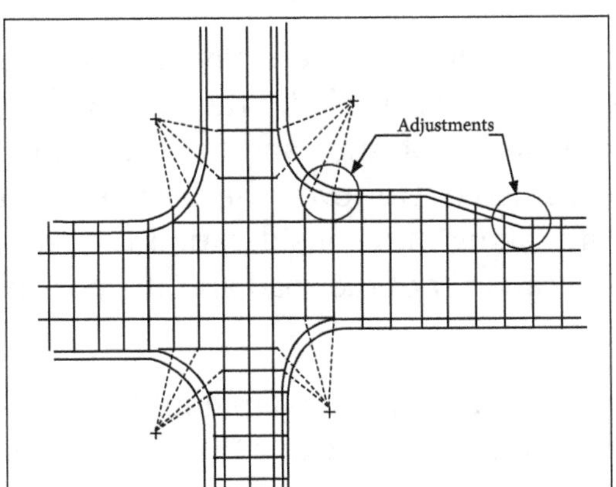

5.3.7 IRC Method

The factors governing design considered are: single and the tandem axle loads, their repetition, tyre pressure and lateral placement characteristics of commercial vehicles.

Wheel Load

Though the legal axle load limits in India have been fixed as 10.2, 19 and 24 tonnes for single axles, tandem axles and tridem axles respectively, a large number of axles operating on National Highways carry much higher loads than legal limits. Data on axle load distribution of the commercial vehicles is required to compute the number of repetitions of single and tandem axles of different weights expected during the design period. For this purpose, an axle load survey may be conducted for a day, covering a minimum sample size of 10 per cent in both the directions. Higher axle loads induce very high stresses in the pavement and result in the consumption of fatigue resistance of concrete. Contribution of the different axle load groups towards fatigue damage must be determined for pavement design. Tyre pressures and shape of contact areas of the commercial vehicles also govern load stresses. For most of the commercial highway vehicles, the tyre pressure ranges from about 0.7 to 1.0 MPa but it is found that stresses in concrete pavements having thickness of 20 cm or more are not affected significantly by the variation of tyre pressure in the range mentioned earlier.

A tyre pressure of 0.8 MPa may be adopted for design. For computation of stresses in the pavements, the magnitude of axle loads must be multiplied by Load Safety Factor (LSF). This takes care of unpredicted heavy truck loads. For important roads, like, the Expressways, National Highways and other Roads where there will be uninterrupted traffic flow and high volumes of truck traffic, the suggested value of LSF is 1.2. For roads of lesser importance having lower proportion of the truck traffic, LSF may be taken as 1.1. For residential and other streets that carry small number of commercial traffic, the LSF may be taken as 1.0. It is recommended that the basic design of the slab be done with a 98th percentile axle load, and the design thereafter checked for fatigue consumption for higher axle loads.

Design Period Normally, the cement concrete pavements have a life span of about 30 years and should be designed for this period. When the traffic intensity cannot be predicted accurately for a long period of time, and for low volume roads, a design period of twenty years may be considered. However, the Design Engineer must use his judgement about the design life taking into consideration the factors, like, traffic volume, the traffic growth rate, the capacity of the road and the possibility of augmentation of capacity.

The Design Traffic Assessment of average traffic should normally be based on seven-thy 24-hour count made in accordance with IRC: 9 "Traffic Census on Non-Urban Roads". The actual value of growth rate 'r' of heavy commercial vehicles should be determined.

However, if actual data is not available, annual growth rate of 7.5 per cent may be adopted. It may be noted that flexural stress caused by axle loads is maximum when the tyre imprint is tangential to the longitudinal edge. When the wheels are tangential to the transverse joints, stresses are lower and when the tyre position is even 15 cm away from the longitudinal edge, there is a significant reduction in the flexural stress. Observation of the lateral distribution characteristics of wheel paths for two-lane two-way roads in India indicates that very few vehicles travel along the edge. A design traffic of 25 per cent of the total two-lane two-way commercial vehicles may be considered as a very conservative estimate for design against fatigue failure. In case of four-lane and multi-lane divided highways, 25 per cent of the total traffic in the direction of predominant traffic may be taken for design of pavement. In case of new highway links, where no traffic count data is available, data from roads of similar classification and importance may be used to predict the design traffic intensity.

The cumulative number of repetitions of axles during the design period may be computed from the following formula:

$$C = \frac{365 \times A\left\{(1+r)^n - 1\right\}}{r}$$

where,

C = Cumulative number of axles during the design period

A = Initial number of axles per day in the year when the road is operational

r = Annual rate of growth of the commercial traffic (expressed in decimals)

n = Design period in years

$$C = \frac{365 \times A\left\{(1+r)^n - 1\right\}}{r}$$

Expected number of applications of different axle load groups during the design period can be estimated from the axle load spectrum.

In most design problems, it is expected that the weights and number of trucks travelling in each direction are fairly equal. This may not be true for roads, such as, haul roads in mine areas where many of the trucks haul full loads in one direction and return empty in other direction. In such cases, a suitable adjustment should be made. It is recommended that the basic design of the slab be done with a 98th percentile axle load, and the design thereafter checked for fatigue consumption for higher axle loads.

Temperature Differential

Temperature differential between the top and bottom. of concrete pavements causes the concrete slab to warp, giving rise to stresses. The temperature differential is a function of solar radiation received by pavement surface at the location, losses due to wind velocity, etc., and thermal diffusivity of concrete, and is thus affected by geographical features of the pavement location. As far as possible, values of the actually anticipated temperature differentials at the location of the pavement should be adopted for the pavement design. For this purpose, guidance may be had from table.

Characteristics of Subgrade and Sub-Base

Strength

The strength of subgrade is expressed in terms of modulus of subgrade reaction k, which is defined as pressure per unit deflection of foundation as determined by plate bearing tests. As the limiting design deflection for the cement concrete pavements is taken as 1.25 mm, the k-value is determined from the pressure sustained at this deflection. As k-value is influenced by test plate diameter, the standard test is to be carried out with a 75 cm diameter plate.

Table: Recommended temperature differentials for concrete slabs.

Zone	States	Temperature Differential, °C in Slabs of thickness			
		15 cm	20 cm	25 cm	30 cm
I	Punjab, U.P., Uttaranchal, Gujarat, Rajasthan, Haryana and North M.P., excluding hilly regions.	12.5	13.1	14.3	15.8
II	Bihar, Jharkhand, West Bengal; Assam and Eastern Orissa, excluding hilly regions and coastal areas.	15.6	16.4	16.6	16.8
III	Maharashtra, Karnataka, South M.P., Chattisgarh, Andhra Pradesh, Western Orissa and North Tamil Nadu, excluding hilly regions and coastal areas.	17.3	19.0	20.3	21.0
IV	Kerala and South Tamil Nadu, excluding hilly regions and coastal areas.	15.0	16.4	17.6	18.1
V	Coastal areas bounded by hills.	14.6	15.8	16.2	17.0
VI	Coastal areas unbounded by hills.	15.5	17.0	19.0	19.2

A frequency of one test per km per lane is recommended for assessment of k-value, unless the foundation changes with respect of subgrade soil, type of sub-base or the nature of formation (i.e., cut or fill) when additional tests may be conducted.

In case of homogeneous foundation, test values obtained with plates of smaller diameter may be converted to the standard 75 cm plate value by experimentally obtained correlations given by,

$$k_{75} = 0.5 \times k_{30}$$

Where, k_{75} and k_{30}, are the k-values obtained on 75 cm and 30 cm diameter plates respectively. Equation is regarded as approximate only. However, in case of layered construction, the tests with smaller plates give greater weightage to the stronger top layer, and the direct conversion to 75 cm plate values by the above correlation somewhat overestimates the foundation strength, and such conversion must be regarded as very approximate only.

The subgrade soil strength and consequently the strength of the foundation as a whole, is affected by its moisture content. The design strength obviously should be the minimum that will be available under the worst moisture conditions encountered. The ideal period for testing subgrade strength would, thus, be during or soon after the monsoon when the subgrade may have attained its highest moisture content.

In case the tests have to be conducted at some other period, especially during the dry part of the year, allowance for loss in subgrade strength due to increase in the moisture must be made. For this purpose, an idea of the expected reduction in strength on saturation of the subgrade may be had from laboratory CBR tests on subgrade soil samples compacted at field density and field moisture content and tested before and after saturation. An approximate idea of k-value of a homogeneous soil subgrade can be obtained from its soaked CBR value using following table. It is advisable to have a filter layer above the subgrade for the drainage of water to prevent (i) excessive softening of subgrade and (ii) erosion of the subgrade particularly under adverse moisture condition.

Table: Approximate k-value corresponding to CBR values for homogeneous soil subgrade.

Soaked CBR value %	2	3	4	5	7	10	15	20	50	100
k-value (kg/cm²/cm)	2.1	2.8	3.5	4.2	4.8	5.5	6.2	6.9	14.0	22.2

The recommendations of IRC:15-2002 shall be followed and if the k-value tested on wet condition of the subgrade is less than 6.0 kg/cm²/cm, cement concrete pavement should not be laid directly over the subgrade. A Dry Lean Concrete (DLC) sub-base is generally recommended for modem concrete pavements, particularly those with high intensity of traffic. The sub-base of DLC should conform to "Guidelines for the Use of Dry Lean Concrete as Sub-base for the Rigid Pavement, IRC:SP:49-1998". In the case of problematic subgrade, such as clayey and expansive soils, etc. appropriate provisions might be made for blanket course in addition to the sub-base as per the relevant stipulations of IRC:15-2002.

The approximate increase in the k-values of subgrade due to different thicknesses of sub-bases made up of untreated granular, cement treated granular and dry lean concrete (DLC) layers may be taken from following tables. 7-day unconfined compressive strength of cement treated granular soil should be a minimum of 2.1 MPa. Dry Lean Concrete should have a minimum compressive strength of 7 MPa at 7 days.

Table: k-values over granular and cement treated sub-bases.

k-value of subgrade, (kg/cm²/cm)	Effective k (kg/cm²/cm) over untreated granular layer sub-base of thickness in cm			Effective k (kg/cm²/cm) over cement treated sub-base of thickness in cm		
	15	22.5	30	10	15	20
2.8	3.9	4.4	5.3	7.6	10.8	14.1
5.6	6.3	7.5	8.8	12.7	17.3	22.5
8.4	9.2	10.2	11.9	--	--	--

Table: k-values over dry lean concrete sub-base.

K-value of subgrade kg/cm²/cm	2.1	2.8	4.2	4.8	5.5	6.2
Effective k over 100 mm DLC, kg/cm²/cm	5.6	9.7	16.6	20.8	27.8	38.9
Effective k over 150 mm DLC, kg/cm²/cm	9.7	13.8	20.8	27.7	41.7	--

The maximum value of effective k shall be 38.9 kg/cm²/cm for 100 mm of DLC and 41.7 kg/cm²/cm for 150 mm of DLC.

Separation Layer between Sub-base and Pavement

Foundation layer below concrete slabs should be smooth to reduce the inter layer friction. A separation membrane of minimum thickness of 125 micron polythene is recommended to reduce the friction (Ref. IRC:15-2002) between concrete slabs and dry lean concrete sub-base (DLC).

Drainage Layer

To facilitate the quick disposal of water that is likely to enter the subgrade, a drainage layer may be provided beneath the pavement throughout road width above the subgrade. The recommendations contained in IRC:15-2002 in this regard may be followed.

Characteristics of Concrete

Design Strength

Since the concrete pavements fail due to bending stresses, it is necessary that their design is based on the flexural strength of concrete. The mix should be so designed that minimum structural strength requirement in the field is met at the desired confidence level. Thus, if

S_1 = Characteristic flexural strength at 28 days

S = Target average flexural strength at 28 days

Z_a = Tolerance factor for the desired confidence level, known as the standard normal variate

σ = Expected standard deviation of field test samples; if it is not known, it may be initially assumed as per 1S:456-2000

Then the target average flexural strength is given as,

$$S = S_1 + Z_a \sigma$$

Table: Values of standard normal variate for different values of tolerance (1S:10262).

Accepted proportion of low results (tolerance)	Quality Level	Standard Normal Variate, Z_a
1 in 15	Fair	1.50
1 in 20	Good	1.65
1 in 40	Very Good	1.96
1 in 100	Excellent	2.33

For pavement construction, the concrete mix should preferably be designed and controlled on the basis of flexural strength. Flexural strength should be determined by modulus of rupture tests under third point loading. The preferred size of the beam should be 15 cm × 15 cm × 70 cm when the size of the aggregate is more than 19 mm. When the maximum size of aggregates is less than 19 mm, 10 cm × 10 cm × 50 cm beams may be used. 1S:516 should be referred to for the test procedure.

Modulus of Elasticity and Poisson's Ratio

The modulus of elasticity, E, and Poisson's ratio, of cement concrete are known to vary with concrete materials and strength. The elastic modulus increases with increase in strength, and Poisson's ratio decreases with increase in the modulus of elasticity. While it is desirable that the values of these parameters are ascertained experimentally for the concrete mix and materials actually to be used in the construction, this information may not always be available at the design stage. Even a 25 per cent variation in E and p values does not have any significant effect on,the flexural stresses in pavement concrete. It is suggested that for design purposes, the following values may be adopted for concrete for the flexural strength of 4.5 MPa.

Modulus of elasticity of concrete E = Experimentally determined value or,

3.0 x 105 kg/cm^2

Poisson's ratio $\mu = 0.15$

Coefficient of Thermal Expansion

The coefficient of thermal expansion of concrete (α) of the same mix proportions varies with the type of aggregate. However, for design purposes, a value of $\alpha = 10\times10\text{-}6$ per °C may be adopted in all cases.

Fatigue Behaviour of Cement Concrete

Due to repeated application of flexural stresses by the traffic loads, a progressive fatigue damage takes place in the cement concrete slab in the form of gradual development of micro-cracks especially when the applied stress in terms of flexural strength of concrete is high. The ratio between flexural stress due to the load and the flexural strength of concrete is termed as the stress ratio (SR). If SR is less than 0.45, the concrete is expected to sustain infinite number of repetitions. As stress ratio increases, the number of load repetitions required to cause cracking decreases. The relation between fatigue life (N) and stress ratio is given as,

$$N = \text{unlimited for } SR < 0.45$$

$$N = \text{unlimited for } SR < 0.45$$

$$N = \left[\frac{4.2577}{SR - 0.4325}\right]^{3.268} \text{ when } 0.45 \le SR \le 0.55$$

$$\text{Log}_{10} N = \frac{0.9718 - SR}{0.0828} \text{ for } SR > 0.55$$

The values of fatigue life for different values of stress ratio are given in table. Use of the fatigue criteria is made on the basis of Miner's hypothesis. Fatigue resistance not consumed by repetitions of one load is available for repetitions of other loads.

Table: Stress ratio and allowable repetitions in cement concrete.

Stress ratio	Allowable Repetitions	Stress Ratio	Allowable Repetitions
0.45	6.279×10^7	0.66	5.83×10^3
0.46	1.4335×10^7	0.67	4.41×10^3
0.47	5.2×10^6	0.68	3.34×10^3
0.48	2.4×10^6	0.69	2531
0.49	1.287×10^5	0.70	1970
0.50	7.62×10^5	0.71	1451
0.51	4.85×10^5	0.72	1099
0.52	3.26×10^5	0.73	832
0.53	2.29×10^5	0.74	630
0.54	1.66×10^5	0.75	477
0.55	1.24×10^5	0.76	361
0.56	9.41×10^4	0.77	274
0.57	7.12×10^4	0.78	207
0.58	5.4×10^4	0.79	157
0.59	4.08×10^4	0.80	119
0.60	3.09×10^4	0.81	90
0.61	2.34×10^4	0.82	68
0.62	1.77×10^4	0.83	52
0.63	1.34×10^4	0.84	39
0.64	1.02×10^4	0.85	30
0.65	7.7×10^3		

5.3.8 Rigid Pavements for Low Volume Roads

The slab of a rigid pavement has significant bending resistance, acting like a bending plate. Vertical deflection of the slab is negligible, and the vertical contact stress on the top of the base is widely spread. In turn, the compressive stress transmitted by the base to the subgrade is dispersed and decreased even more, to the point where vertical displacement of the subgrade is negligible. Faulting at slab joints is avoided by specifying tongue and groove edges and/or dowels.

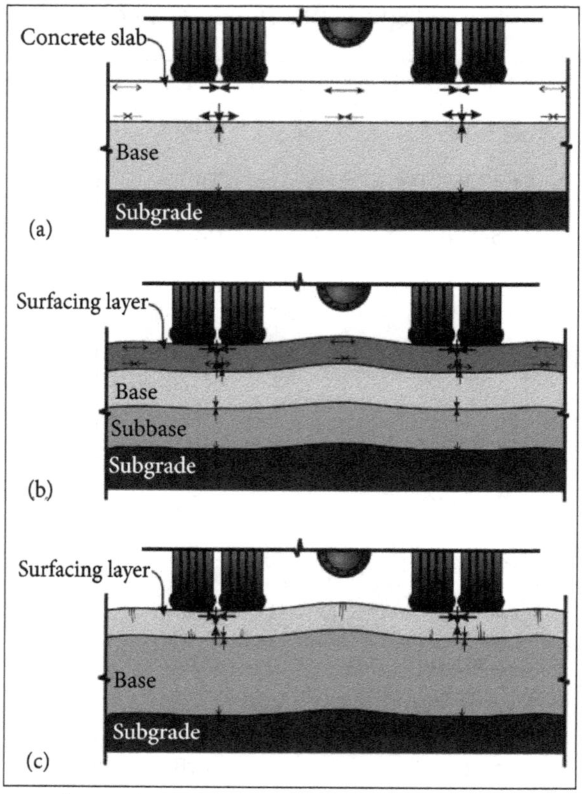

Load-carrying mechanisms for (a) rigid pavements. (b) flexible asphalt pavements, and (c) flexible granular pavements.

Arrow direction and weight indicate compression or tension, and relative magnitude of the stresses. Note the bending stresses in the slab of the rigid pavement, the weaker bending stresses in the asphalt pavement, and the lack of tension in the granular pavement. Cracking in the granular pavement is shown on the top surface outside the dual wheels and in the bottom of the surfacing layer under the dual wheels. Note that for simplicity, the crossfall has not been shown.

Factors Governing Design

Wheel Load

Heavy vehicles are not expected frequently on rural roads. The maximum legal load limit on single axle with dual wheels in India being 100 kN, the recommended design load on dual wheel is 50 kN having a spacing of the wheels as 310 mm centre-to-centre. Agricultural tractors and trailers also are being used to carry construction material and the single wheel load may rarely approach 50 kN.

Tyre Pressure

The tyre pressure may be taken as 0.8 MPa for a truck carrying a dual wheel load of 50 kN while for a wheel of tractor trailer; the pressure may be taken as 0.5 MPa. The effect of tyre pressure on the wheel load stresses for practical thickness of pavement is not significant.

Design Period

Concrete pavements designed and constructed as per the guidelines contained in this document will have a design life of 20 years or higher, as evidenced from the performance of roads constructed in the past in the country. The design methodology given in these guidelines is based on wheel load stresses. The repetitions of axle loads, curling stresses and the consumption of fatigue for different axle loads, which form the basis of design in IRC:58, need not be considered for low volume traffic except in special situations where heavy truck traffic is anticipated.

Design Traffic for Thickness Evaluation

For traffic less than 50 CVPD, only wheel load stresses for a load of 50 kN on dual wheel need be considered for thickness estimation since there is a low probability of maximum wheel load and highest temperature differential between the top and the bottom of the rigid pavement occurring at same time. For traffic higher than 50 and less than 150 CVPD, thickness evaluation should be done on the basis of the total stresses resulting from wheel load of 50 kN and temperature differential. For traffic exceeding 150 CVPD, fatigue could be a real problem and the guidelines consider thickness evaluation on the basis of fatigue fracture considering a reliability of 60 percent against a reliability of 90 percent adopted for IRC:58.

40 percent of the pavement slabs are expected to crack at the end of the design period against 10 percent cracking considered for the high volume roads. For the fatigue analysis of a concrete pavement the cumulative number of commercial traffic at the end of the design period can be estimated from the following formula.

$$N = A \times \left(\frac{(1+r)^n - 1}{r} \right) \times 365$$

where,

A = Initial CVPD after the completion of the road

r = rate of traffic increase in decimal (for 5 percent rate of increase in traffic, r = 0.05)

n = design period in years (recommended as 20 years)

N = total number of cumulative commercial vehicles at the end of the design period

An estimation of gross weight of vehicle and axle load can be made from the knowledge of goods that the vehicles carry. Approximate number of commercial vehicles having an axle load of about 100 kN as a percentage of total number of commercial vehicles from above equation may be estimated from the traffic survey. For A > 150 CVPD, a default value of ten percent of the values obtained from Equation 3.1 may be considered for cumulative fatigue analysis as illustrated later. In case the road is expected to carry a large volume of very heavy trucks on a regular basis, spectrum of axle loads may have to be considered to avoid damage to pavements and design concept of IRC:58 may be used.

Characteristics of the Subgrade

The strength of subgrade is expressed in terms of modulus of subgrade reaction, k, which is determined by carrying out a plate bearing test, using 750 mm diameter plate according to 18:9214-1974. In case of homogeneous foundation, test values obtained with a plate of 300 mm diameter, k_{300} may be converted to give k_{750} determined using the standard 750 mm dia. plate by the following correlation:

$$k_{750} = 0.5\, k_{300}$$

Since, the subgrade strength is affected by the moisture content, it is desirable to determine it soon after the monsoon. Stresses in a concrete pavement are not very sensitive to minor variation in k values and hence its value for a homogeneous soil subgrade may be obtained from its soaked CBR value using the below table. It can also be estimated from Dynamic Cone Penetrometer also as described in IRC:58.

Table: Approximate k value corresponding to CBR values.

Soaked subgrade CBR	2	3	4	5	7	10	15	20	50
k Value (MPa/m)	21	28	35	42	48	50	62	69	140

The minimum CBR of the subgrade shall be 4.

Sub-base

A good quality compacted foundation layer provided below a concrete pavement is commonly termed as subbase. It must be of good quality so as not to undergo a large settlement under repeated wheel load to prevent cracking of slabs.

The provision of a sub-base below concrete pavement has many advantages such as:

i. It provides a uniform and reasonably firm support

ii. It supports the construction traffic even if the subgrade is wet

iii. It prevents mud-pumping of subgrade of clays and silts

iv. It acts as a leveling course on distorted, non-uniform and undulating sub-grade

v. It acts as a capillary cut-off

Types of Sub-base

i. Traffic up to 50 CVPD

75 mm thick compacted Water Bound Macadam Grade III (WBM lll)/Wet Mix Macadam (WMM) may be provided over 1 00 mm granular subbase made up of gravel, murrum or river bed material with CBR not less than 30 percent, liquid limit less than 25 percent and Plasticity Index less (PI) less than 6. If aggregates are not available within a reasonable cost, 150 mm of cement/lime/lime-fly ash treated marginal aggregate/ soil layer with minimum Unconfined Strength (UCS) of 3 MPa at 7 days with cement or at 28 days with lime/lime-flyash may be used. The stabilized soil should not erode as determined from wetting and drying test (IRC:SP:89).

ii. Traffic from 50 to 150 CVPD

75 mm thick WBM lll/WMM layer over 1 00 mm of granular material may be used as a subbase. Alternatively, 100 mm thick cementitious granular layer with a minimum unconfined strength (UCS) of 3 MPa at 7days with cement or 28 days with lime/lime-flyash over 100 mm thick cementitious naturally available materials with a minimum UCS of 1.5 MPa with cement at 7 days or with lime or lime-flyash at 28 days may be provided.

iii. Traffic from 150 to 450 CVPD

150mm thick WBM lll/WMM over 100 mm of granular subbase may also be used. Alternately, 100mm of cementitious granular layer with a minimum UCS of 3.0 MPa at 7 days with cement or at 28 days with lime or lime-flyash over 1 00 mm of the cementious layer with naturally occurring material with a minimum UCS of 1.5 MPa at 7 days with cement or at 28 days with lime or lime-flyash. The cementitious marginal aggregates may be much cheaper than WBM/WMM in many regions having acute scarcity of aggregates.

The granular subbase and WBM layers should meet the requirement of MORD Specifications, Section 400(34). Quality of subbases varies from region to region and past experience on performance of concrete pavements in different regions is tlie best guide for the selection of the most appropriate subbases.

Commercially available IRC accredited stabilizers with no harmful leachate also may be used if found successful on trials.

Effective Modulus of Subgrade Reaction over Granular and Cement Treated Subbases

For the granular subbases, the effective k value may be taken as 20 percent more than the k value of the sub-grade shown in the above table. for the cementitious subbases, the effective k value may be taken as twice that of the subgrade.

Recommendations for estimated of effective modulus of subgrade reaction over granular or cemented subbase are given in the below table. Reduction in stresses in pavement slab due to higher subgrade CBR is marginal since only fourth root of k matters in stress computation but the loss of support due to erosion of the poor quality foundation below the pavement slab under wet condition may damage it seriously.

The GSB layer with fines passing 75 micron sieve less than 2 percent can act as a good drainage layer and addition of 2 percent cement by weight of total aggregate will make it non-erodible. Most low volume roads with the concrete pavements in built up area having WBM over GSB have performed well even under adverse drainage conditions.

Table: Effective k values over granular and cementitious subbases.

Soaked CBR	2	3	4	5	7	10	15	20	50
k Value over granular subbase (thickness 150 to 250mm), MPa/m	25	34	42	50	58	60	74	83	170
k Value over 150 to 200 mm cementations sub-base, MPa/m	42	56	70	84	96	100	124	138	280

It should be ensured that embankment, the subgrade and the subbase shall be well compacted as per MORD specifications (34) otherwise heavy wheel loads may displace subbase under adverse moisture condion leading to cracking of the unsupported concrete slab.

Concrete Strength

Since concrete pavements fail due to bending stresses, it is necessary that their design is based on the flexural strength of concrete. Where there are no facilities for determining the flexural strength, the mix design may be carried out using the compressive strength values and the following relationship:

$$f_f = 0.7 \sqrt{f_{ck}}$$

where,

f_f = flexural strength, MPa

f_{ck} = characteristic compressive cube strength, MPa

For Low volume roads, it is suggested that the 90 day strength may be used for design since concrete keeps on gaining strength with time. The 90 day flexural strength may be taken as 1.10 times the 28 day flexural strength or as determined from laboratory tests. 90 day compressive strength is 20 percent higher than 28 day compressive strength. Heavy traffic may be allowed after 28 days.

The concrete mix must be so designed that the minimum flexural strength requirement in the field is met at desired confidence level. For rural roads, the tolerance level can be taken as 1 in 20. The normal variate, Z_a, for this tolerance level being 1.65, the target average flexural strength is obtained from the following relationship:

$$S = S_1 + Z_a \sigma$$

where,

S = target average flexural strength, at 28 days, MPa

S_1 = characteristic flexural strength, at 28 days, MPa

Z_a = normal variate, having a value of 1.65, for a tolerance factor of 1 in 20

σ = expected standard deviation of field test samples, MPa

Table: Expected Values of Standard Deviation of Compressive Strength.

Grade of Concrete	Standard Deviation for Different Degrees of Control, MPa		
	Very Good	Good	Fair
M 30	5.0	6.0	7.0
M 35	5.3	6.3	7.3
M 40	5.6	6.6	7.6

For pavement construction for rural roads, it is recommended that the characteristic 28 day compressive strength should be at least 30 MPa and corresponding flexural strength shall not be less than 3.8 MPa.

Modulus of Elasticity and Poisson's Ratio

The Modulus of Elasticity, E, of concrete and the Poisson's ratio may be taken as 30,000 MPa and 0.15 respectively.

Coefficient of Thermal Expansion

The coefficient of thermal expansion of concrete, α may be taken as,

$$\alpha = 10 \times 10^{-6} \text{ per } ^\circ\text{C}$$

Fatigue Behaviour of Concrete Pavement

For most rural roads, fatigue behavior is not important because of low volume of commercial vehicles. In case a rural road forms a connecting link between two important roads or if the road connects several villages, there can be significant amount of traffic consisting of buses and trucks due to the agriculture, construction and social activities, and fatigue behavior of pavement slab may be considered in such cases. Fatigue equations of IRC:58 cannot be used because they are valid for 90 percent reliability. Following fatigue equation for 60 percent reliability is recommended rural village roads.

$$\log_{10} N_f = \frac{SR^{-2.222}}{0.523}$$

Where, N_f is fatigue life of a pavement subjected to stresses caused by the combined effect of wheel load of 50 kN and temperature gradient. If the number of heavy vehicles are large, fatigue analysis should be done for the spectrum axle load recommended in IRC:58.

Influence of the light commercial vehicles is negligible. Occasional heavy loads may not affect the pavements since the subgrade is weak only during the certain period in the monsoons or post monsoon periods and only worst value of subgrade strength is considered in design. Concrete also keeps on gaining strength with time. SR can be defined as,

$$SR = \frac{\text{flexural stress due to wheel load and temperature}}{\text{flexural strength}}$$

5.4 Continuously Reinforced Cement Concrete Pavements

Continuously Reinforced Concrete Pavement (CRCP) is constructed with steel reinforcing bars placed within the concrete along entire length of the pavement. CRCP naturally forms tight transverse cracks to evenly transfer loads.

The reinforcing bars control the width of the transverse cracks that form and hold them closed. The transverse cracks do not impair the structural integrity of the pavement.

The result is a continuous, smooth-riding surface capable of withstanding heaviest traffic loads and the most adverse environmental conditions. Because of its longer life expectancy, greater durability, and minimal maintenance requirements, CRCP can provide the best long-term value of any pavement type.

It was first developed more than 75 years ago, Continuously Reinforced Concrete Pavement (CRCP) became popular with the construction of the U.S. Interstate Highway system in the 1960s and 1970s. Many stretches of highway which are constructed at that time are still in use, having outperformed and outlived their original loading and design-life projections.

Today, CRCP has become the material of choice in the U.S. and in countries around the world for the construction that includes highways, airport runways, roadways, railway track beds, and warehouse flooring, because CRCP means:

- Less impact on the environment because it uses 100-percent recycled material and reduces effect of thermal-heat absorption in densely populated metro areas.
- A smooth riding surface that contributes significantly to a greater rider comfort and increased fuel efficiency.

- Lower cost to owners because it requires less maintenance.
- Fewer delays for the motorists because it virtually eliminates reconstruction and repairs.

Construction of CRCP

Construction of CRCP is similar to that of other concrete pavement types. Planning and execution are crucial since errors made during these stages may be detrimental to the overall success of the project. It is important to pay special attention to certain details such as the selection and installation of the reinforcement, carrying out of the construction joints, and so forth. As with any other type of pavement, the base should be finished to ensure a uniform roadbed for the reinforcement supports and construction equipment as well as to provide with a uniform slab thickness.

The base must ensure proper drainage to the slab base interface and be non-erodible to limit the potential of punch-outs. A permeable base fully satisfies these criteria. First, transverse reinforcement bars are manually placed on the metal supports by teams of steel fixers. A sufficient amount of supports will prevent any collapsing under a 250 kg load. Their design should be in accordance with the concrete cover specifications.

The longitudinal reinforcement bars are placed on transverse ones and then tied to the latter. Generally, it is recommended that longitudinal reinforcement be placed on the upper third section of the slab to limit crack openings. A sufficient amount of the concrete cover above the reinforcement is necessary to prevent any corrosion.

A minimum spacing of 150 mm between reinforcement bars is recommended to ensure adequate steel cover. The longitudinal bars may be welded with one another or tied. If tied, the recommended overlap is 25 to 35 bar diameters. The overlaps are usually offset from one lane to the next to ensure they are not all in the same cross section. The free ends of CRCP are exposed to movements mainly caused by temperature differentials.

Systems are installed at each end to restrict the movements from the last 100 meters of the slab. Tie bars should be placed in longitudinal construction joints to keep slab edges together on either side the joint. Special attention must be paid when forming the transverse construction joints when concrete placement is completed at end of the day.

The phases subsequent to the placement of CRCP (finishing, texturing, curing, saw cutting of the longitudinal joints and sealing) resemble to that of other slab types.

Performance of the CRCP

Its main objectives are to improve pavement life and performance as well as to optimize the use of the funds allocated to the construction and maintenance of the road network.

The survey includes:

- Distress mapping on the 150 m sections and general survey of the entire CRCP projects.
- Measurements of crack openings and end joints.
- Measurements of longitudinal profile.
- Measurements of transversal profile.
- Coring and sampling.
- Measurements of deflections on the slab and at the joint edges.
- Measurements of skid resistance and macro texture.
- Measurements of salt penetration levels in the concrete.
- Measurements of steel corrosion potential.

5.4.1 Roller-Compacted Concrete Pavements

Roller-compacted concrete pavement is a zero slump concrete that is placed with standard or high-density paving equipment and consolidated/compacted using steel-drum or rubber-tired rollers to achieve a durable, wear resistant surface.

Roller-compacted concrete pavements were first used in the 1970's for stabilization of logging roads in the US and Canada. It has proven to be very effective heavy duty pavement that can stand up to harsh climates, heavy wheel loads and the difficult operating conditions.

Typically, roller-compacted concrete has been used for heavy duty pavements like the log handling yards, intermodal terminals, freight depots, and other industrial applications. However, more recently there has been an increase in the use of roller-compacted concrete to create cost-effective pavements for many conventional highway and street applications.

The high strength of RCC pavements eliminates common and costly problems traditionally associated with asphalt pavements.

RCC pavements are:

- Span soft localized subgrades
- Resist rutting
- Do not deteriorate from spills of fuels and hydraulic fluids
- Will not deform under heavy, concentrated loads
- Will not soften under high temperatures

Benefits of RCC pavements are:

- Durability
- Speed of Construction
- Competitive Cost
- Low Maintenance
- Sustainability

6

Highway Construction and Maintenance

6.1 Types of Highway Construction

Highway Construction

Highway construction involves all aspects of the building process beginning with clearing of the native soil, preparation of the surface, placement of the pavement material, and preparation of the final roadway for use by traffic. Highways initially were built with manual labor assisted by horse-drawn equipment for grading and moving materials. Today, modern construction equipment is used for clearing the site, grading the surface, compaction of the pavement base courses, transporting materials, and placing the final highway pavement.

Advances in construction equipment have made the rapid building of large highway sections possible. Nuclear devices test com-paction of soil and base courses, Global Positioning Systems (GPS) and Geographic Information Systems (GIS) are used to establish line and grade, and specialized equipment for handling concrete and bridge work are all innovations in the construction industry. Large, automatically controlled mix plants have been constructed, and new techniques for improving durability of structures and the substitutions for scarce materials have been developed.

6.1.1 Earthwork

Construction of building foundation, road construction, and construction of sewer lines, water supply lines involve the job of cutting and filling of earth. This job is called "Earthwork".

Shoring and Timbering

Length, breadth and depth of trench should be as per the drawing and design given in the map. Where soil is wet or cohesion less the side slopes are supported by the use of cross bars and side shutters (planks) in order to prevent collapse of the trench. This is called shoring and timbering. The details of shoring and timbering areshown in the below figure.

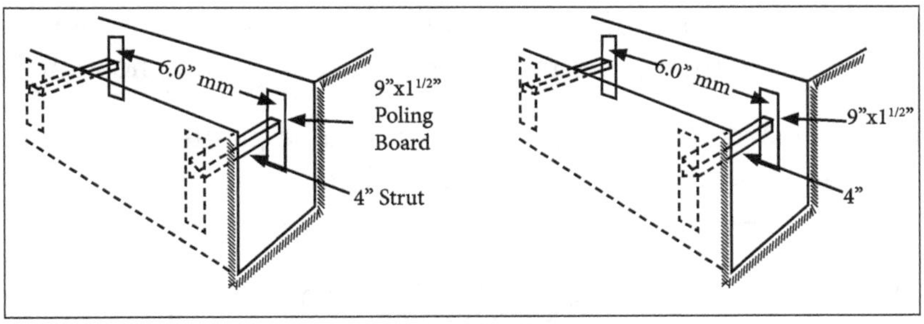

Loose timbering

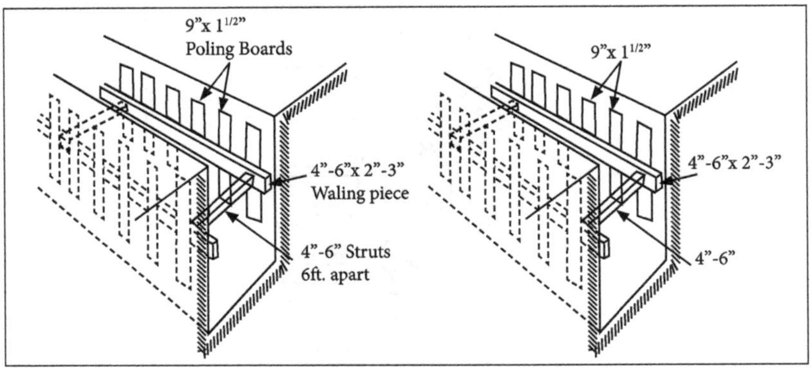

Close timbering

Unit of Measurement of Earthwork

The unit of measurement of the earthwork is cubic meter (m³). If the length of trench = 1 m, breadth 1 m and depth 1 m then total earthwork is 1 m × 1 m × 1 m = 1 cub m. (length × breadth × depth).

If the shape of the trench is trapezoidal,

$$\text{Vol. of earth work} = \frac{B_1 + B_2}{2} \times \text{height} \times \text{length}$$

Base of the Trench

The base of the trench must be made flat and all loose soil should be removed from it. If by mistake the depth of trench becomes more than the required depth, it should not be filled by loose soil rather it should be filled with the lean concrete 1 : 4 : 8 or 1 : 5 : 10 and made flat under the guidance of the site engineer.

If it is required to construct a soil embankment, minimum of 10% more soil should be added i.e. if the designed height of the embankment is 6.0 m, the embankment of height 6.6 m should be constructed initially to allow for the settlement of soil at a later stage.

In case of deep filling or cutting a side slope of 1 : 1 or 2 : 1 (i.e., 2 horizontal and 1 vertical) is usually maintained. If it is required to put new soil over old soil, Benching is required to be done so that a perfect bonding between old and new soil is achieved.

Benching means construction of steps. Grass should be planted on the slope to check the soil erosion and to protect the slope of the embankment.

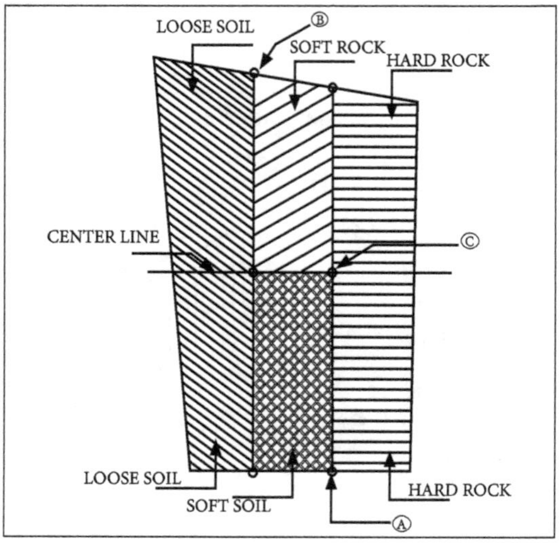

Apart from the above earthwork is required to be carried out in the construction of roads. During construction of roads two types of situation can arise i.e., it may require either cutting or filling. During excavation we may get different types of soil layers. For this a reformed line is to be made for controlling the width and direction of the road/ embankment. This helps in determining whether the filling or cutting has been done as per the drawing or not. Catch water drains are made on the sides of the cutting which takes out water from the trench.

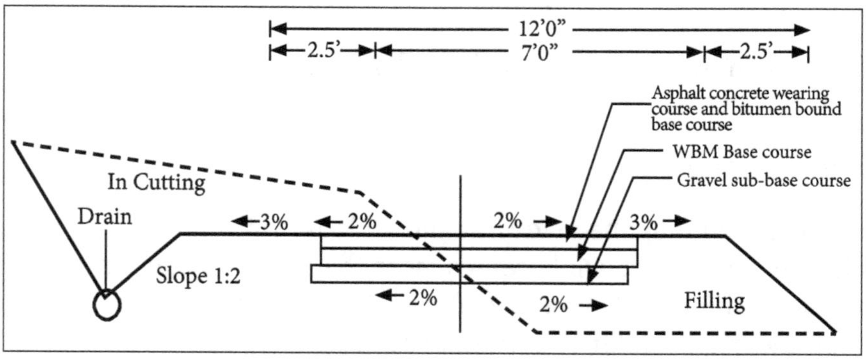

Cross section of a road.

Payment

The rate of payment for cutting or filling of earth is made by cubic m of the earth work and the lead distance for carriage of earth away from the site or to site is also taken into account.

6.1.2 Construction of Earth Roads

Earth roads are constructed as detailed below:

Materials Required

The materials required for the construction of an ordinary earth road are natural soil, originally available at site, a suitable stabilizer, if required. The ideal composition of soil which is recommended for construction earth roads.

Method of Construction of an Ordinary Earth Road

The construction work of an ordinary earth road is done in the following steps:

1. Preparing the subgrade: The subgrade is prepared and the surface is brought to required camber and gradient.
2. Rolling an watering: The surface is rolled properly and well soaked with water.
3. Spreading the soil: After watering, a layer of graded soil, about 10 cm thick is evenly spread.
4. Rolling: The layer of soil is rolled at an optimum moisture content (OMC) with a suitable roller and finally finished with a light roller.

   The type of the roller for compaction is decided based on soil type, desired amount of compaction and availability of the equipment. Atleast 95 per cent of dry density of its light compaction is considered desirable. Sometimes, a second layer (around 10 cm thick) is also spread and rolled properly to act as wearing course.
5. Checking: The camber of the finished surface is checked and corrected, if necessary.
6. Curing: The rolled surface is watered for upto four or five days for curing. No traffic should be allowed to pass over the road surface during this period.
7. Opening to traffic: After curing, The compacted earth surface is allowed to dry out for few days and then is opened to traffic.

Method of Construction of a Stabilized Earth Road

The construction work of a stabilized earth road is completed in steps as explained in case of ordinary earth road. For constructing a stabilized earth road, the subgrade is prepared to the required camber and gradient. Then the base course is prepared by one of the suitable method of soil stabilized soil (about 10 cm thick) is spread over the base course and is rolled properly. The surface is then cured and opened to traffic.

6.1.3 Gravel Roads

A gravel road is a type of the unpaved road surfaced with gravel that has been brought to the site from a quarry or stream bed. They are common in the less developed nations, and also in the rural areas of developed nations such as Canada and the United States.

Gravel Road Construction

Although gravel road construction can be done anytime on any surface, a smooth, long-lasting roadway requires the right process. Like any other building project, a gravel road starts with the foundation. The other three stages are the sub-grade material, sub-base material and gravel surface.

Weak areas are potential problems waiting to happen. Professional contractors will survey the driveway or proposed project, inspect the road or plan, anticipate problems, and will suggest workable solutions. With the help of an experienced road builder, the result will be a solid, gravel road that is offering comfortable passage for years to come.

Four Parts of Gravel Road Construction

Driver satisfaction is almost guaranteed when we use the right material and design. The depth of the road base depends on the ground; the hard packed dirt and rock require little or no base, while sandy or swampy areas need special attention. This foundation supports road, and is usually comprised of hard-packed dirt.

The sub-grade layer is composed of larger rocks, to give it solidarity and strength.

The sub-base layer of smaller rocks helps to fill in the gaps in the sub-grade layer, and whole thing is packed down to create a solid foundation.

Gravel, of course, is put on the top. The experienced road contractors are particular about this topcoat.

A knowledgeable contractor should know two things: how to build the road and how to fix problems.

The four most common complaints are springtime softness, summer/fall drainage problems, rutting, and dust. Although the best time to remedy the first two problems is during initial construction, if the current gravel road is sagging or soggy, we can add a stronger base into the existing road, or redirect water flow.

If the ridges in the road are slight and erode over time, the road needs to be graded. However, if these ruts are deep and occur after a rain or snowstorm, the roadbed is too soft and requires more material. As roads get drier, dust is a common nuisance.

Property owners have two choices in dealing with this problem: drive slower, or have the road treated with a surface stabilizer.

6.1.4 Water Bound Macadam Roads

Water Bound Macadam road consists of coarse aggregate mechanically interlocked by rolling and bonded together with screening, binding material and water lain on a prepared sub-base or on existing pavement as the case may be.

Binding is achieved by using stone dust as filler with water. Water bound Macadam may be used as a sub-base, base course or surfacing wait.

The aggregate may be one of the following :

(i) Crushed stone

i. Broken stone

ii. Over-burnt brick aggregate

iii. Crushed slag

iv. Natural aggregate such as kankar or laterite

Table: Grading required for coarse aggregate.

No.	Grading size range	Seive designation	% by weight passing the sieve
1.	90 mm to 40 mm Jhama Brick aggregate	100 mm 80 mm 63 mm 40 mm 20 mm	100 65-85 25-60 0-15 0-5
2.	40 mm to 63 mm soft stone, Lime stone, flint, kankar, quartzite, laterite and brick ballast	80 mm 63 mm 50 mm 40 mm 20 mm	100 90-100 35-70 0-10 0-5
3.	50 mm to 20 mm hard stone Granite, trap, basalt Diorite etc.	63 mm 50 mm 40 mm 20 mm 10 mm	100 95-100 35-70 0-10 0-5

Wbm Road Construction Procedure

Construction of water bound macadam road involves the following six basic steps as given below:

1. Preparation of sub-grade
2. Preparation of sub-base

3. Preparation of base
4. Preparation of wearing course
5. Preparation of shoulders
6. Opening to traffic

Step 1: Construction of the Sub-Grade

Sub-grade act as a cushion for other layers i.e. in order to achieve durable road sub-grade must be strong. Sub-grade is provided by digging up the sub-soil and the level of the sub-grade is decided by subtracting total thickness of the pavement from the finished level of the road pavement.

The sub-grade is thoroughly compacted by the rollers weighing 8 tonnes by sprinkling water one night before. Low spots which develop during rolling should be made up and brought to the grades as required. In rocky regions, the sub-grades are not rolled whereas in region of clay soils, a layer to natural sand, moorum or gravel, is provided over sub-grade and is duly packed.

Step 2: Construction of the Sub-Base

On a well compacted sub-grade, spread 10 to 20 cm size boulders or broken stones, or over burnt bricks in layers of 15 cm thickness and total width of the sub-base to be kept 60 cm wider than pavement width, projecting around 30cm on each side. The sub-base should be compacted by a roller to provide an even surface.

Step 3: Construction of the Base

On prepared sub-base or directly on the sub-grade, as the case may be, the specified materials of the base course is spread and a proper grade, thickness and cross-sections maintained as per design shown on the supplied drawings.

Step 4: Preparation of the Wearing Course

This course may be laid in one or two layers according to total designed thickness and the thickness of each layer should not exceed 10 cm. this component being very important, the following steps may be taken systematically.

Check the defective portions/patches of the newly laid base course and rectify them.

Provide either bricks on end edging or earthen kerbs strong enough to prevent the new road material from spreading outward and also to retain the water used in consolidation of the wearing course.

Spread road metal evenly over the prepared base to the specified thickness and hand pack them so that the finished surface is brought to the required camber.

Spread coarse aggregate over the surface and roll it dry with a suitable roller till interlocking of aggregate is achieved with sufficient void space. The rolling is started from the edges and gradually shifted towards centre.

After dry rolling, spread the screening materials (stones upto 12 mm size) with uniform rate so that voids of coarse aggregates get filled properly. This is achieved by the dry rolling and brooming alternatively, till the voids of the coarse aggregates are filled.

After spreading screening material, sprinkle sufficient quantity of water, sweep the surface and roll it with roller again.

Now apply binding material in two to three thick layers at a slow and uniform rate. Each layer of binding material is rolled after the addition of sufficient water. The slurry is swept in with brooms to fill the void properly. The moving wheel of the roller must be cleaned with water. Continue the operations of spreading of binder, sprinkling of water, sweeping with the brooms and rolling till the voids get filled and slurry forms a wave before the moving wheel of the roller.

After the proper compacting allow it to dry over night. Spread a layer of sand or earth, about 6 mm thick and roll the surface again after sprinkling the water lightly.

The surface may be allowed for 7 to 10 days of curing.

Step 5: Construction of the Shoulders

While curing the pavement surface, prepare the shoulders by filling earth to the specified cross slope and compact them properly by tamping or by rolling. Width and thickness of the shoulder should be as per specification.

Step 6: Opening to the Traffic

After properly drying, the road pavement may now be opened to traffic, ensuring that the traffic is distributed uniformly over the full width of the pavement.

6.1.5 Bituminous Pavements and Construction of Cement Concrete Pavements

Construction Procedure of Pavement Slab

Sequence of construction of the pavement slab consists of the following steps:

1. Preparation of subgrade and base

2. Placing of forms
3. Installation of joints
4. Batching of aggregates and cement
5. Mixing and placing concrete
6. Consolidation and finishing concrete
7. Cutting of concrete

1. Preparation of Subgrade and Base

The subgrade or base should be prepared complying with the following conditions:

i. No soft spots are present in the subgrade or base.

ii. Subgrade or base should be uniformly compacted and extended about 30 cm, on either side of the width of pavement to be concreted.

iii. Subgrade or base should be adequately drained.

iv. Plate load test conducted on the subgrade should yield a minimum modulus of subgrade reaction of 5.5 kg/cm^3.

Steps to be followed before placing concrete:

i. The subgrade is prepared and checked atleast two days in advance of concreting.

ii. The subgrade or base is kept in moist condition at the time of concrete so as to avoid the subgrade absorbing water from concrete. If necessary, the subgrade may be saturated well with water for 6 to 20 hours prior to placing of concrete.

iii. Water proof paper may be spread on subgrade if concrete to be placed directly on the subgrade. In this case no saturation of subgrade is necessary.

2. Placing of Forms

Wooden or steel forms are used. Wooden forms have minimum base width of 10 cm for 20 cm slab thickness and of 15 cm for slabs over 20 cm thickness. Forms are joined neatly and are set with exact grade and alignment. Forms are rigidly fixed such that during the entire operation of concreting they should not deviate more than 3 mm from straight edge of 3 m length.

Steel forms commonly used are straight 3 m sections. They are aligned vertically and horizontally by slip joints and held in position by three or more steel stakes. These stakes are of 2.5 cm diameter, 45 to 75 cm long driven at intervals at back of form. Forms are available in heights of 15 to 30 cm and with corresponding base width over a similar range. These are made of 4 to 6 mm steel plate.

3. Installation of Joints

Extreme care should be taken in all operations connected with joints. Face of transverse joints should be straight, perpendicular to the center line of pavement and also perpendicular to the surface of the finished slab. Load transfer devices like dowel bars used in expansion joint should be aligned and placed accurately. There should be free movement of slab ends in longitudinal direction.

4. Batching of Aggregates and Cement

Based on the design concrete mix, the proportion of ingredients like coarse aggregate and fine aggregate are proportioned by weight in a weight batching plant. These are placed in the hopper of the mixer along with the necessary quantity of cement.

Cement is measured by the bag which measures 50 kg. All batching of material is done on the basis of one or more whole bags of cement taking the unit weight of cement as 1440 kg/m^3. The mixing is done in batch mixer.

Mixer performs well such that there is uniform distribution of materials and mixed material is uniform in color and homogeneous. The batch of cement, fine aggregate and coarse aggregates are added to the mixer. The water for mixing is introduced into the drum within the first 15 seconds of mixing. After all the materials are placed in the mixer, the mixing is to be started within one and a half minutes.

5. Mixing and Placing Concrete

The ingredients are mixed in required quantity for immediate use and is deposited on the subgrade or base. The concrete should be deposited to the correct depth and width of pavement section within the formwork. The operation of placing concrete should be continuous. Care should be taken to see that no seggregation of materials takes place in any stage of the operation.

6. Consolidation and Finishing

Concrete is spread uniformly by shovels with re-distribution wherever needed. Needle vibrator is used for compaction. Surface of the pavement is compacted either using a power-driven finishing machine or using a vibrating hard screen. Hard compaction may be resorted to where the width of slab is very small and at corners.

After placing and operating concrete, the crown and the cross-section of the pavement has to be done with the required grade. Concrete is further compacted parallel to carriageway center line by using a longitudinal float.

After the longitudinal floating is done and the excess water gets disappeared the final grade and level are tested with the straight edge. Before the concrete gets hardened, the surface is belted with a two-ply canvas belt.

After belting, the pavement is given a broom finish with fiber broom brush. Before the concrete develops initial set, the edges of the slab are finished with edging tool.

7. Curing of Concrete

It is very important to ensure proper curing of the finished concrete.

Following are the methods usually adopted:

i. Earth cover kept wet

ii. Hay or straw cover kept wet

iii. Cover of wet felt mats or cotton mats

iv. Sawdust kept wet

v. Continuous sprinkling of water

vi. Surface application of calcium chloride or sodium sulphate

vii. Cover by water-proof paper or impervious membrane

Forms are not removed until concrete has set for atleast 12 hours. Special devices are used to pull the forms without damaging the pavement. At the end of curing period, the surface grooves in longitudinal and transverse joints are cleaned and filled with sealing compound. After the clearing the road surface, the road is open to traffic.

Bituminous Road Constructions Steps

1. Preparation of the Existing Base Course Layer

The existing surface is prepared by removing the pot holes or rust if any. The irregularities are filled in with the premix chippings at least a week before laying surface course. If existing pavement is extremely way, a bituminous leveling course of adequate thickness is provided to lay a bituminous concrete surface course on a binder course instead of directly laying it on a WBM.

2. Application of Tuck Coat

It is desirable to lay an AC layer over a bituminous base or binder course. A Tack coat of bitumen is applied at 6.0 to 7.5 kg per 10 sq. m area, this quantity can increase up to 7.5 to 10 kg for non-bituminous base.

3. Preparation and Placing of Premix

Premix is prepared in a hot mix plant of a required capacity with the desired quality control. The bitumen can be heated upto 150 – 177 °C and the aggregate temperature should not differ by over 14 °C from the binder temperature.

The hot mixed material is accumulated from the mixture by the transporters, carried to the location is spread by a mechanical paver at a temperature of around 121 to 163 °C. The camber and the thickness of the layer are accurately verified. The control of temperatures during the mixing and the compaction are of great significance in the strength of resulting pavement structure.

4. Rolling

A mix after it is placed on the base course is thoroughly compacted by rolling at a speed up to 5 km per hour.

The initial or break down rolling is done by 8 to 12 tonnes roller and intermediate rolling is done with a fixed wheel pneumatic roller of 15 to 30 tonnes having a tyre pressure of 7 kg per sq. cm. the wheels of the roller are kept damp with water.

The number of passes required depends on thickness of the layer. In warm weather rolling on the next day, helps to increase the density if the initial rolling was not adequate. The final rolling or the finishing is done by 8 to 10 tonne tandem roller.

Tandem roller

5. Quality Control of the Bituminous Concrete Construction

The routine checks are carried out at site to ensure the quality of the resulting pavement mixture and pavement surface.

Periodical checks are made for:

a. Aggregate grading

b. Temperature of aggregate

c. Grade of bitumen

d. Temperature of paving mix during mixing and compaction

At least one sample for every 100 tonnes of the mix discharged by the hot mix plant is collected and tested for above requirements. Marshall tests are conducted. For every 100 sq. m of the compacted surface, one test of the field density is conducted to check whether it has a minimum of 95% of the density obtained in the laboratory. The variation in the thickness allowed is 6mm per 4.5 m length of construction.

6. Finished Surface

Road construction-finishing.

The AC surface should be checked by a 3.0 m straight edge. Longitudinal undulations should not exceed 8.0 mm and the number of undulations higher than 6.0 mm should not exceed 10 in a length of 300 m. The cross-traffic profile should not have undulations exceeding 4.0 mm.

Concrete Pavements

Concrete pavements are rigid pavements having very high flexure strength as compared to flexible pavements. Concrete pavements can be constructed using two different methods:

1. Alternate bay method
2. Continuous bay method

In alternate bay method, concrete pavement slab are laid on whole width of pavement in alternate bays.

In the continuous bay method, the concrete pavement slabs are laid continuously only on one bay and another bay is open for the traffic.

Usually, the second method of continuous bay is preferred over alternate bay method because, traffic movement is allowed while in the first it is restricted. Also, the alternate empty spaces invites the rainwater collection and create in-convenience to construction work.

Various steps for the construction of concrete pavements:

1. Preparation of the sub-grade and sub-base
2. Placing of the forms
3. Batching of the material and mixing
4. Transporting and placing of the concrete
5. Compaction and finishing
6. Floating and the straight edging
7. Belting, brooming and edging
8. Curing of the cement concrete

Subgrade Preparation

The subgrade preparation involves cleaning, earthwork and compaction.

Where the concrete layer is laid directly over the subgrade, subgrade is moist at the time concrete is placed. If the subgrade is dry, water may be sprinkled over the surface of the subgrade before laying any concrete course, however, care should taken so that soft patches or water pools are not formed at surface (IRC:15-2002, Chakroborty and Das 2003).

As an alternative arrangement, concreting may be done over a water proof polyethylene sheet, and in that case moistening the subgrade surface becomes redundant. The polyethylene sheet acts as a capillary cut-off layer (IRC:15 2002). Figure below presents a photograph of subgrade construction in progress.

Preparation of subgrade.

Construction of Base/Sub-base

A base/sub-base to the concrete pavement provides a uniform and reasonably firm support, prevents the mud-pumping, and acts as capillary cut-off. Sub-base for concrete pavement may be constituted with brick flat soling, WBM, granular aggregates, crushed concrete, slag, stabilized soil etc.

As per IRC:15 (2002), the sub-base could be of three types with:

i. Granular material (for example, brick soling with one layer of sand under it, WBM, the well graded granular materials etc.)

ii. Stabilized soil

iii. Semi-rigid material, (for example, lime fly ash concrete, lime burnt clay puzzolana concrete, lean cement concrete roller-compacted concrete etc.)

Dry Lean Cement Concrete as Sub-base

The thickness of DLC, generally recommended is 100mm or 150mm (IRC:SP-49 1998). The maximum aggregate to cement ratio is 15:1. The average compressive strength of DLC cubes at 7 days, as recommended by the Indian guidelines (IRC:SP-49 1998) should not be less than 10 MPa, tested on 5 samples and individual compressive strength must not be less than 7.5 MPa, at 7 days (MORT&H 2001, IRC:SP-49 1998).

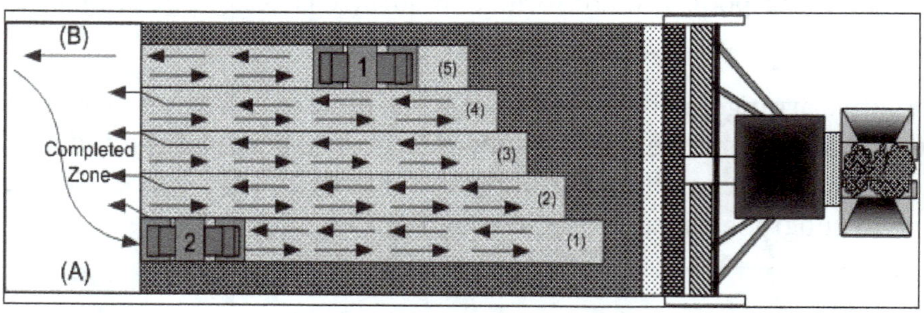

Compaction sequence for DLC.

Here, (A) is the outer edge, and (B) is the inner edge of the pavement

Laying of DLC

Before construction of DLC sub-base, the prepared subgrade is sprinkled with water to moisten the surface. The material is to be laid uniformly by a paver without any segregation. The paving machine should have high amplitude paving bars. The curing of DLC can be done by spraying a liquid curing compound, or by covering the surface by gunny bags. As per Indian guidelines, the construction of cement concrete pavement can only start after 7 days of DLC construction (MORTH 2001, IRC:SP-49 1998).

Laying of polyethylene sheet over DLC.

Compaction and Surface Finishing

Concrete is spread evenly and is rodded with suitable equipment such that formation of honey-combing or voids can be avoided. At the same time, over-compacting must be avoided, which may cause segregation and loss of entrained air (Swanlund and Vanikar 2002). The working of the fixed form and slip-form pavers are different, as explained below.

Fixed Form Paving System

In fixed form paver system, there are separate powered machines for spreading, compacting and finishing are used. The spreader spreads concrete evenly through reversible auger to the desired surcharge level (O'Flaherty 2002). The rotary strike-off paddles trim slight irregularities in the surface of the surcharge concrete and adjusts with the carriage-way cross-slope. Compaction beam applies vibration to the concrete with pre-designed amplitude and the frequency (O'Flaherty 2002). This vibration also helps to put the dowel and tie bars at their desired positions.

The wet formed joint groove is made by introducing a vertical cut immediately after compaction is over and inserting a preformed cellular permanent strip. As an alternative, a saw joint groove can be made after the concrete is adequately hardened and can maintain a sharp edge itself (O'Flaherty 2002).

The finishing of the surface is made, usually, with a pair of finishing beams. The leading beam vibrates and smoothens surface, and the rear beam acts as float. Beams are oriented obliquely so that it causes a less damage to the joints (O'Flaherty 2002).

Slip-form Paving System

Slip-form paving machine is a self-propelled system which can automatically spread, trim, compact and finish the surface in a synchronized manner through its feedback sensors. Placing of dowel or tie bars at their pre-designed locations are done by slip-form pavers. The introduction of the joint grooves, surfacing texturing and spraying of curing compound etc. are done by the equipment those follow the paver (O'Flaherty 2002).

Slip-form paver requires a few guide-wires, parallel to edge of construction and maintained at a fixed height, installed on the both side. The alignment of slip-form paver is controlled automatically with respect to the guide-wires. Correct and precise alignment of the guide-wires is therefore extremely important.

The hopper/spreader maintains a constant surcharge of the concrete above conforming plate level. The conforming plate, vibrators, strike off paddles and the finishing screed gives the final shape of the concrete pavement (O'Flaherty 2002). Figure below explains schematically the operation of a typical slip-form paver, and Figure below a photograph of concrete pavement construction by a slip-form paver.

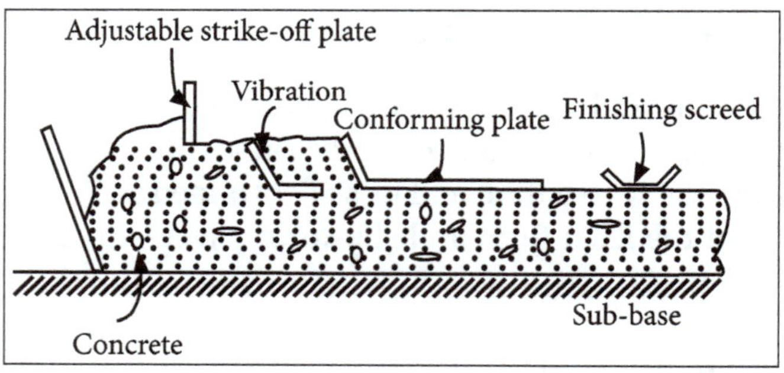

Schematic diagram explaining operation of a typical slip-form paver.

Construction of concrete pavement slab with slip-form paver.

Texturing

Finished concrete has a smooth surface; texturing of concrete surface is done to impart required skid resistance to the concrete surface. The texturing is done by means of wire brushing or grooving along the transverse direction. Initial texturing may be done at the time of construction of the paver itself. Final texturing is done no sooner the sheen of the concrete surface goes off (Swanlund and Vanikar 2002).

Texturing of finished surface by texturing machine.

Concrete Curing

Curing is a process in which requisite moisture content and temperature is maintained so that concrete achieves its design strength through hydration of cement. For initial curing, curing compound with high water retentivity may be spread over the finished surface to prevent rapid drying of water.

For final curing, moistened hessain/gunny bags or continuous ponding should be kept for about a fortnight. As an alternative arrangement to ponding, impervious liquid maybe spread over the surface so as to restrict evaporation of water from laid concrete. Forms are removed from the freshly prepared concrete layer after about curing of fourteen hours (IRC:15 2002, Chakroborty and Das 2003).

Curing of finished concrete pavement.

Opening to Traffic

After the curing period is over and before opening the road to traffic, the temporary seal material is to be removed, and the joints are to filled with the recommended joint sealing compound. The pouring of sealing material is monitored carefully such that it is not spilled over the pavement surface.

Construction of joints and joint sealing have been discussed separately (IRC:15 2002, Chakorborty and Das 2003).

Recapitulation

The construction of concrete pavement involves construction of subgrade, sub-base/ base and the concrete slab.

The subgrade preparation involves cleaning, earthwork and compaction.

The various steps involved in cemented/concrete layer construction, depending on requirements, are spreading, trimming, compacting, jointing, finishing, curing, texturing etc. The construction cemented/ concrete is generally done by automated equipment, called paver.

Pavers are of two types, fixed form and slip-form. Slip-form paving machine is a self-propelled system that can perform paving in a synchronized manner through its feedback sensors.

6.2 Pavement Failures

Causes of distress in flexible pavements:

It may be attributed to settlement of any one of the components of the pavement.

Failure in Subgrade Soil

The main cause for failure of flexible pavement is the excessive settlement of the subgrade of soil. Figure shows the settlement of the subgrade and the effects on the base and wearing course. This can be noticed in the form of excessive undulations or waves and corrugations on the pavement surface. The failure of subgrade may be attributed to inadequate stability and excessive stress application.

Subgrade stability is due to excessive moisture, improper compaction and inherent weakness of the soil. Stability of subgrade is the resistance to settlement under the short and long term stress conditions. If the load transferred on to the subgrade due to wheel loads is less then the deformation of the subgrade will be less and will be in the elastic range.

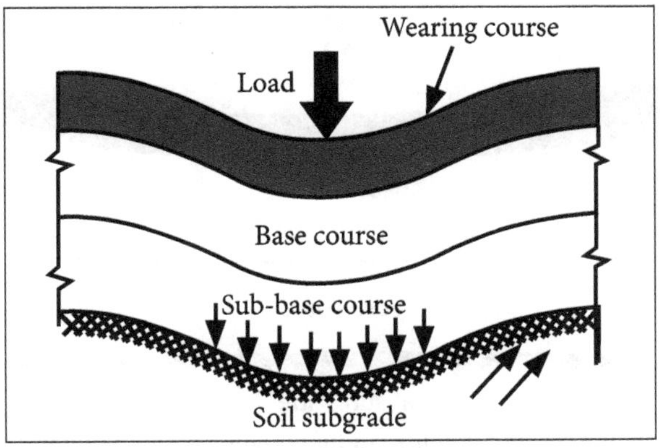

Failure of subgrade.

Continued stress application causes the deformation to be plastic and unrecoverable resulting in subsidence of the subgrade. Figure shows the elastic and plastic deformations of a pavement.

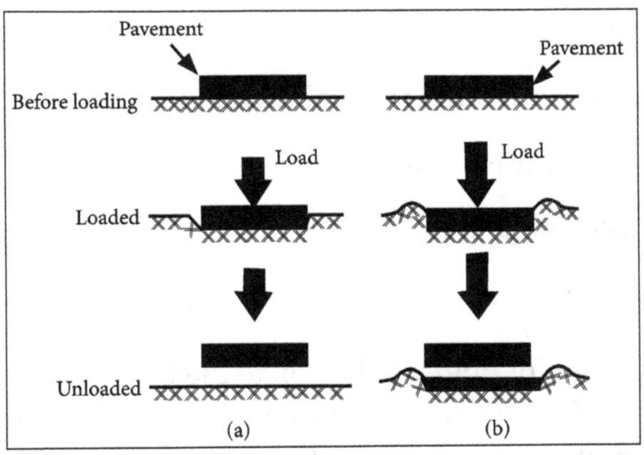

Deformation of pavement: (a) Elastic deformation (b) Plastic deformation

Failures of Sub-base or Base Course

Sub-base and base courses may settle due to internal readjustment of aggregates and movement of wheel loads may result in loosening of the compacted layer accounting for disturbance and loss of binding action. Thus deformation would occur.

Inadequate thickness or poor mix of base or sub-base courses may lack stability. Improper quality control during construction of sub-base or base course may cause settlement.

Loss of base course materials is also possible when either the base course is not covered properly or has completely worn out. Due to moving vehicles, a sort of suction is caused which leads to removal of binding material in WBM base.

The exposed aggregates may form dust due to abrasion and attrition and loose stone aggregates forming a pot hole.

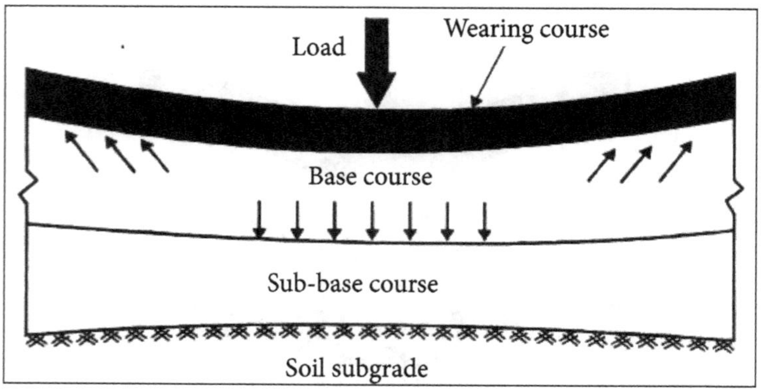

Failure of base course.

Failure of Wearing or Surface Course

Inadequate thickness or absence of wearing course results in damage of base course immediately. The damaging effect is more due to variation in climatic conditions particularly in rain or frost weather. Example: Bullock carts traffic.

Failure of wearing surfaces are due to the following:

i. Inferior or improper mix design

ii. Inadequate binder cement

iii. Inferior quality of binder result in a poor bituminous surfacing

Volatilisation and oxidation of binders also make the bituminous surfacing brittle. Because of this chemical action, cracking of the pavement occurs which allows seepage of rain water to harm the underlying layers.

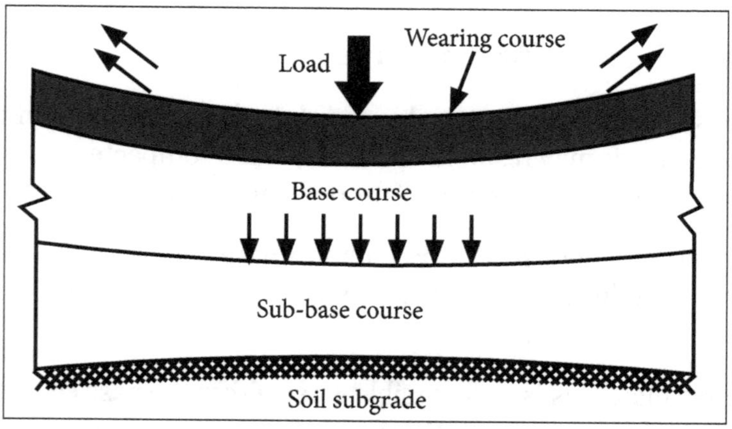

Failure of wearing course.

Failure due to Inferior Materials and Construction

Many failures are attributed to the use of inferior materials in the pavement. Some materials may appear to satisfy initially but may show rapid deterioration. Continued weather change further deteriorates all the layers resulting in cracking.

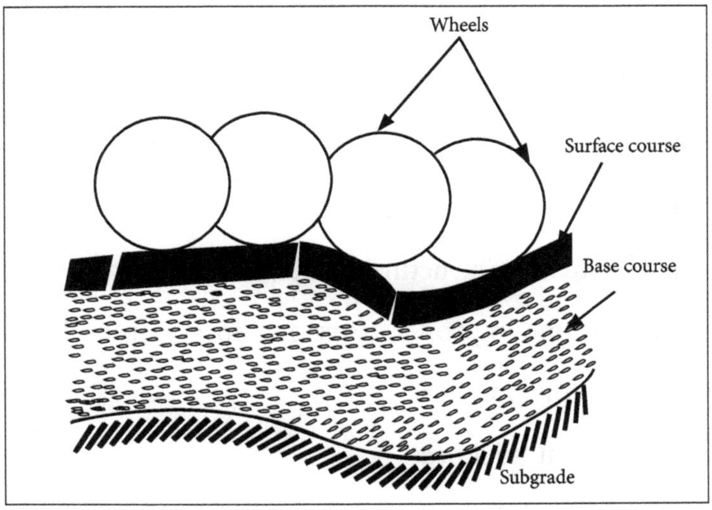

Failure due to improper compaction of subgrade and pavement layer.

Improper construction of layers and inadequate quality control may yield failure of one or all the layers of the pavement. Figure shows failure of base course, surface course and subgrade because of improper compaction.

Causes of Distress of Rigid Pavements

Distress or failure in cement concrete pavements may be attributed to three factors as detailed below:

1. Deficiencies in pavement materials
2. Structural inadequacy
3. Improper construction and maintenance

Deficiencies in Pavement Materials

Failures or distress in the cement concrete pavements are due to improper selection of materials and their quality.

Following are the chief causes:

1. Soft aggregates
2. Dirty aggregates with silt and clay
3. Low quality joint filler
4. Poor sealer material

5. Poor quality steel
6. Improper use of cement for the specific region

Structural Inadequacy

Non-collection of proper traffic data, improper assessment of future traffic demand, improper design approach, improper assessment of subgrade strength, climatic condition, etc are attributed to the structural inadequacy.

Following are the main causes of structure inadequacy:

1. Poor subgrade soil and improper assessment of its strength
2. Improper mix design approach
3. Inadequate pavement thickness
4. Incorrect spacing of joints
5. Incorrect design of load transfer devices
6. Absence of longitudinal hinge joints
7. Long length of slab
8. Non-existence of temperature steel
9. Deep foundation movements

Improper Construction and Maintenance

Many of the failures or distress are due to the following factors:

1. Poor workmanship in pavement and joint construction
2. Poor surface finish
3. Improper and insufficient curing
4. Use of concrete mixes which are wet
5. Over finishing at the edges
6. Improper alignment of dowel bar and lubrication of bars
7. Infiltration of incompressible materials (like sand) into the joint
8. Unplanned annual maintenance
9. Special maintenance after adverse weather conditions

6.3 Maintenance of Highways

Maintenance operations of highways fall under three categories noted below:

1. Routine maintenance
2. Special repairs
3. Resurfacing

1. Routine Maintenance

All types of pavements are exposed to climatic changes and adverse weather conditions, evidently these would wear out. To avoid further deterioration and to reduce further heavy expenses, it is mandatory to maintain the highway in time.

The following works are to be maintained routinely:

i. Upkeeping of carriageway

ii. Shoulders

iii. Subgrade

iv. Side drain clearance for silt accumulation

v. Highway geometric and traffic control

vi. Ancillary works like culverts, bridges, etc.

2. Special Repairs

Adverse weather conditions and other causes may demand special repairs of the pavement which would have been damaged. Need may arise for special repair of some sub-surface drainage system and cross drainage structure.

3. Resurfacing

It arises due to the expiry of the life of the pavement or surface being heavily damaged.

6.4 Pavement Evaluation

A study of various factors pertaining to pavement, such as subgrade support, pavement composition and its thickness, traffic loading and environmental condition, is referred to as pavement evaluation. Aim of pavement evaluation is to assess as to whether and to what extent the pavement fulfills the design requirements.

Based on the evaluation, the maintenance and strengthening works could be planned. Study is used to investigate the structural adequacy of pavements and also the requirements for providing safe and comfortable traffic operations.

Although several methods are used, the four methods categorized below are the basic ones:

1. Structural Evaluation of pavements
2. Evaluation of pavement surface condition
3. Present serviceability index
4. Visual Rating

Roughness of Surface

Apart from different aspects, the riding quality of a pavement is a function of the specifications adopted for the surfacing. Roughness is one among them. A measure of the pavement performance can be obtained by monitoring its roughness.

Smooth riding surface brings about many advantages, viz., higher speeds, less fuel consumption, less wear and tear of tyre, less consumption of spare parts, riding comfort and safety. In performance evaluation, rigidity of surface play a vital role.

Present Serviceability Index

A rating system was developed by AASHO involving the measurement of permanent deformation, riding quality and the extent of cracking and patching. Rating is well-known by the term Present Serviceability Index (PSI). It is the most widely used pavement rating measure.

Following equations give the value of PSI for flexible and rigid pavements,

1. Flexible Pavements

$$ISI = 5.03 - 1.91\log(1+SV) - 1.38(RD)^2 - 0.01\sqrt{C+P}$$

2. Rigid Pavements

$$ISI = 5.41 - 1.80\log(1+SV) - 0.09\sqrt{C+P}$$

where,

SV = Slope variance giving an index of the longitudinal profile

RD = Rut depth under a 4 ft. straight edge

C = Cracking distress in terms of lineal feet of cracks per 1000 sq. ft. area

P = Patched area in sq. ft. per 1000 sq. ft. area

Skid Resistance

Though smooth surface has various advantages it is dangerous to traffic, especially when the surface is wet and the vehicles move fast. It is normal to construct reasonably skid-resistance surfaces. Passage of vehicles polish the aggregates and excess Bitumen tends to fatten the surface and render it slippery.

Hence it is essential to evaluate the skin resistance of the surface at periodic intervals to ensure that the roughness level has not fallen to dangerous low levels. It is accomplished by periodic measurement of skin-resistance. Skin resistance is measured using towed fifth wheel pump integrator, car-mounted roughness measuring device, etc.

Structural Evaluation of Pavements

Plate bearing test can be conducted for both flexible and rigid pavements to assess the structural capacity. Assessment may be made by the load carried at a specified deflection at a place or by the amount of deflection at a specified load on the plate. Performance of a flexible pavement is closely related to the elastic deformation under loads or its rebound deformation.

Under the existing conditions of a pavement, the deflection of pavement under the design wheel loads serves as an index to carry traffic loads. Benkelman Beam is the most commonly used method.

Before conducting the Benkelman Beam measurements, a rating survey of the road is performed by dividing the road into homogeneous sections of approximately similar serviceability.

Deflection Measurement

It occurs due to various factors such as subgrade soil type, moisture content and compaction level of subgrade, pavement thickness and drainage conditions, pavement surface temperature and wheel load.

Measured using a suitable equipment such as Benkelman Beam, Lacroix Deflectograph, Dynaflect and Deflectome.

6.4.1 Strengthening of Existing Pavements

A highway is expected to have adequate stability to withstand the design traffic under prevailing climatic and subgrade conditions. If the design load and the anticipated traffic are maintained, the planned maintenance will be fruitful.

If the pavement is to support increased wheel loads and load repetitions, the pavements cannot be maintained properly but undergo large distress. Only solution to manage the increased traffic is either to direct the traffic on some adjacent roads or to strengthen the existing pavements.

Strengthening may be done by providing additional thickness of pavement provided the subgrade is strong enough. Construction of one or more layers over the existing pavement is called overlay. If the distress of the road is excess, construction of overlay will not help. The road has to be reconstructed.

Types of Overlays

An overlay may be either flexible or rigid. Flexible overlays are generally bituminous surfaces. It can be constructed over existing flexible pavements or existing concrete pavements. A bond breaker comprising of a granular base is used between a concrete base pavement and overlay. Deflection is also reduced to some tolerable level.

Rigid overlays may consist of plain, simply reinforced or continuously reinforced concrete pavements. Provision of a bond breaker may or may not be necessary. Following are the combination of overlays:

Existing pavement	Overlay
Cement concrete	Cement concrete
Cement concrete	Bituminous
Bituminous or flexible	Cement concrete
Bituminous or flexible	Flexible or Bituminous

Choice of overlay depends on various factors including total thickness of overlay required, wheel load, subgrade strength, etc.

Permissions

All chapters in this book are published with permission under the Creative Commons Attribution Share Alike License or equivalent. Every chapter published in this book has been scrutinized by our experts. Their significance has been extensively debated. The topics covered herein carry significant information for a comprehensive understanding. They may even be implemented as practical applications or may be referred to as a beginning point for further studies.

We would like to thank the editorial team for lending their expertise to make the book truly unique. They have played a crucial role in the development of this book. Without their invaluable contributions this book wouldn't have been possible. They have made vital efforts to compile up to date information on the varied aspects of this subject to make this book a valuable addition to the collection of many professionals and students.

This book was conceptualized with the vision of imparting up-to-date and integrated information in this field. To ensure the same, a matchless editorial board was set up. Every individual on the board went through rigorous rounds of assessment to prove their worth. After which they invested a large part of their time researching and compiling the most relevant data for our readers.

The editorial board has been involved in producing this book since its inception. They have spent rigorous hours researching and exploring the diverse topics which have resulted in the successful publishing of this book. They have passed on their knowledge of decades through this book. To expedite this challenging task, the publisher supported the team at every step. A small team of assistant editors was also appointed to further simplify the editing procedure and attain best results for the readers.

Apart from the editorial board, the designing team has also invested a significant amount of their time in understanding the subject and creating the most relevant covers. They scrutinized every image to scout for the most suitable representation of the subject and create an appropriate cover for the book.

The publishing team has been an ardent support to the editorial, designing and production team. Their endless efforts to recruit the best for this project, has resulted in the accomplishment of this book. They are a veteran in the field of academics and their pool of knowledge is as vast as their experience in printing. Their expertise and guidance has proved useful at every step. Their uncompromising quality standards have made this book an exceptional effort. Their encouragement from time to time has been an inspiration for everyone.

The publisher and the editorial board hope that this book will prove to be a valuable piece of knowledge for students, practitioners and scholars across the globe.

Index

www.ingramcontent.com/pod-product-compliance
Lightning Source LLC
LaVergne TN
LVHW080638210926
848485LV00003B/292